Sunset Homeowner's Guide to Solar Heating

By the Editors
of Sunset Books
and Sunset Magazine

**Lane Publishing Co.,
Menlo Park, California**

The solar scene ... today and tomorrow

With each passing month and year, solar heating (and cooling) arouses increased curiosity, anticipation, and—for many of us—excitement. It promises to bring abundant warmth to our homes from that ultimate renewable resource, the sun's radiant energy. It offers a key approach to a solution of our national energy problems. A fast-growing industry, it is changing at breathless speed from a dream of the future into a present-day reality.

In all the ferment of activity and invention that makes up the current solar scene, we've observed differences of opinion over the most basic principles of solar home heating, owing to the still-evolving state of the art. When we asked a panel of solar experts to check our manuscript,

even they expressed a broad spectrum of opinion on topics ranging from general concepts to minute details. This book represents our efforts to pin down the prevailing concepts and present them in a clear, nontechnical fashion, though with time, new developments may supersede some of the information contained on the following pages.

We'd like to extend our thanks particularly to our checkers— Donald W. Aitken, Director, Center for Solar Energy Applications, San Jose State University; J. Douglas Balcomb, Solar Division, Los Alamos Scientific Laboratory; Francis de Winter, President, Altas Corporation and Chairman, American Section, International Solar Energy Society; Marshall Hunt, Development Division, California Energy Commission; Kathy Lefferts; William Seavy, Solar Coordinator, Pacific Gas and Electric Company; Wayne E. Shannon, Manager, Energy Programs, Lockheed Palo Alto Research Laboratory; Larry Sherwood, New Mexico Solar Energy Association; William A. Shurcliff, Honorary Research Associate in Physics, Harvard University; Donald Watson, AIA; Bruce A. Wilcox, Berkeley Solar Group; David Wright, AIA; John I. Yellott, Distinguished Visiting Professor of Architecture, Arizona State University—for their help in straightening out the knotty details, and to the owners and designers of the homes that appear in this book.

Edited by Holly Lyman Antolini

Special Consultant: Fred Nelson
Associate Editor,
Sunset Magazine

Design: Joe di Chiarro

Illustrations: Mark Pechenik, AIA

Cover: Gathering solar energy, the active water trickle-type collector on the roof heats this home, with some help from the sun-catching south windows. The trellis shades the windows in summer. Design: Interactive Resources. Photographed by Norman A. Plate.

Editor, Sunset Books: David E. Clark

Second Printing January 1979

Contents

Harnessing the Sun

An introduction to basic types of solar heating

There's nothing new under the sun—*solar heating existed when man emerged from the caves and built his first home. Today we're rediscovering how to use the sun's beneficent warmth to heat our homes.*

The use of the sun to heat a home is a daring new concept—and as old as the hills. Before the industrial age, people displayed considerable ingenuity in designing dwellings to benefit from the sun's generous warmth.

Though the Southwestern American Indians didn't call it "solar heating," they nestled their buildings together around south-facing courtyards for a good reason: so that the winter sun's low beams would strike and warm their thick adobe walls, providing natural heating. These Indians had a minutely accurate knowledge of the sun's path, and they took maximum advantage of it in laying out their buildings. Even the thickness of the walls was designed to allow the sun's heat to be stored during the day, until needed at night.

Today's solar designs return to many of the same concepts—orientation of the building toward the south for maximum exposure to winter sun, walls that retain heat, and windows concentrated on the south wall.

But this reveals a new attitude evolving among many Americans today. After years of designing homes that formed artificial climates, using inexpensive and seemingly plentiful fossil fuels to power furnaces and air conditioners, we are beginning to consider our energy resources too precious for use in low-intensity home heating and cooling. We are once again designing homes to take advantage of the sun and other climatic factors, sometimes with a modest sacrifice of comfort and convenience.

In the excitement of rediscovery and new technology, the sun's potential is sometimes overrated. Solar energy is not the simple solution to our energy problems it is sometimes claimed to be. Solar experts still disagree about which heating systems work best and which are most cost-effective. Even to operate a solar heating system can still require great technical know-how. And though we dream of free heat, while the sun showers us daily with many times more energy than we need, it is still a complex and sometimes prohibitively expensive task to design, produce, install, and maintain the equipment needed to capture that solar energy and store it as heat. Even when the task is accomplished, only a rare solar heating system can provide 100 percent of a house's heat needs.

Nevertheless, as homeowners watch their fuel costs rise and grow increasingly concerned about the scarcity of gas, oil, and electricity, many ask urgently, "How will I heat my home in the future?" Expensive and experimental as it presently seems, solar heating is an ever more appealing alternative—and one that has already spurred many to action.

Conservation Comes First

The sun is a capricious heater: here today when the sky is blue, it may be gone tomorrow behind a storm cloud. Solar radiation is low in concentration and intermittent, often disappearing when it's most needed. As one scientist put it, the sun is "a part-time performer in a full-time world." The challenge in solar heater design is to capture, accumulate, and store enough of the sun's sporadic energy to supply heat during the nights and stormy days between sunny periods.

In meeting this challenge, your home itself forms the foundation of good solar design. If it is drafty and uninsulated, it will never contain the sun's low-intensity heat long enough to make solar heating practical. All solar experts agree that, prosaic though it may seem, the rule of thumb is: minimize your house's heat loss first, before you consider any further solar equipment or design.

How Does Your House Lose Heat?

Heat moves naturally from warmer areas to colder ones, seeking to balance the temperatures. It flows through your house, moving from its source toward the coldest areas. If the air outside your house is colder than the air inside, your inside heat will seek any possible route by which to continue its flow to the cold air outside. (When it's hotter outside than in, hot air from outside will flow into the house's cooler interior.)

Ghostlike, heat can pass not only from the air in one room to the air in another, or from object to object, but also right through solid walls. It has three ways of moving from one spot to another: conduction, radiation, and convection. Because they apply to all heat loss or gain, these three

concepts are fundamental to an understanding not only of heat conservation, but also of the operation of all solar heating systems.

Conduction. By passing from molecule to neighboring molecule, heat can move through a solid object or from one object to an adjoining one, provided the two are touching. This is called **conduction.** It is by conduction that heat escapes through a solid uninsulated house wall. Generally, the denser the material of the wall or object, the more quickly the heat can move: concrete, for example, is a better heat conductor than wood, which is less dense.

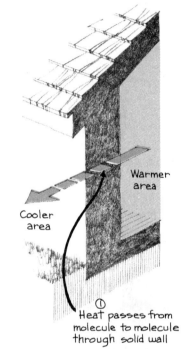

Cooler area

Warmer area

① Heat passes from molecule to molecule through solid wall

Drawing 1: Conduction

Radiation. Like light, heat is transmitted by electromagnetic wave motion. This kind of heat transfer is called **radiation.** Through radiation, heat can jump from warm objects to cooler ones without warming the air in between. The cool object absorbs some of the energy radiated by a warmer object and reflects the remainder. It is by radiation that the sun can warm your wall even on a frosty day (see Drawing 2, page 6).

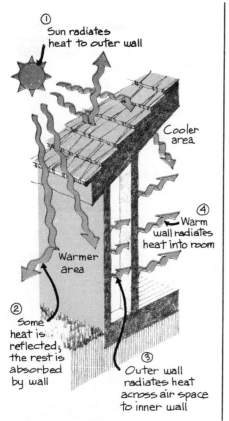

① Sun radiates heat to outer wall

Cooler area

④ Warm wall radiates heat into room

Warmer area

② Some heat is reflected; the rest is absorbed by wall

③ Outer wall radiates heat across air space to inner wall

Drawing 2: Radiation

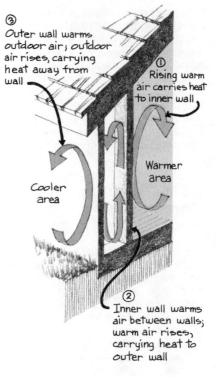

③ Outer wall warms outdoor air; outdoor air rises, carrying heat away from wall

① Rising warm air carries heat to inner wall

Warmer area

Cooler area

② Inner wall warms air between walls; warm air rises, carrying heat to outer wall

Drawing 3: Convection

Convection. Air motion can carry heat from warmer surfaces to cooler ones. This is called **convection.** As air warms, it expands, becomes lighter, and rises. Then, as it gives off heat to surrounding objects, it cools, contracts, becomes denser, and sinks again. In an enclosed space such as a room, heat forms convective currents of rising warm air and falling cool air. When such currents develop between the structural framing members of your wall or roof, quite a lot of heat can pass out from the interior wall or ceiling to the exterior wall or roof (or vice versa, when the outside temperature is high).

Only when air has room enough to move and establish currents, however, does convection take place. If trapped in small spaces or cavities such as those in a sweater's knit or a piece of polystyrene, air becomes a good insulator, helping to prevent the transfer of heat.

Where the Heat Escapes

Your house has many potential trouble spots, where heat can be lost in winter or gained in summer. The first culprits are your walls, roof, and floors: these, if uninsulated, can rob you of two-thirds of your heat through conduction, radiation, and convection. You also lose heat because of air infiltration through cracks and joints, around window and door frames, in interior walls between heated and unheated areas, and in exterior walls near the foundation or wherever two different materials meet in your home's construction.

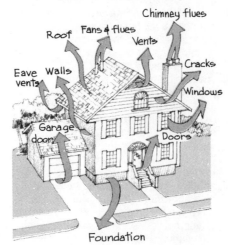

Chimney flues

Fans & flues

Roof

Vents

Cracks

Eave vents

Walls

Windows

Garage door

Doors

Foundation

Drawing 4: Heat loss trouble spots

The next culprits are your windows and doors. Though windows generally make up less of your home's total surface area than walls or roof, they lose much more heat per square foot, especially if they are single glazed, containing only one thickness of glass. Windows and doors together may steal as much as one-third of your heat.

Finally, you stand to lose some heat through deliberate openings, such as vents or chimneys, and from uninsulated heating ducts, water heaters, and hot-water pipes.

Hold That Heat!

Blocking these escape routes is your simplest, most cost-effective energy strategy. Proper insulation, weather-stripping, caulking, and storm doors and windows (or double-glazed windows) constitute the basic requirements for an intended solar house. These measures are the house's barricades against heat loss (or summer heat gain). They are described in detail, and accompanied by do-it-yourself installation instructions, in the *Sunset* book *Do-It-Yourself Insulation and Weatherstripping*.

En Route to Solar: Climate Design

Adapting your home to its specific climate, so that it forms its own protective barriers against heat loss or gain, is an important step toward solar heating. The better a house is assimilated into its climate, the more it tends to maintain its own natural balance of comfort, winter or summer, and the less the sun will have to work to warm it.

Your first move toward solar heating can be as simple as planting a tree to give summer shade to a window, or as ambitious as adding an enclosed porch to shield your front door from winter drafts. Go one step further, and you can design your home to turn its back to the wintry wind and open its walls with windows to the sun's radiant warmth on a chilly December day.

This is designing *with* your climate, reaping its benefits and fending off its punishments. Once you've adopted this approach, you're already well on your way to solar heating—itself one of the major adaptations a house can make to its climate.

Climate and Microclimate

The first step in designing your climate-responsive house (or adapting your existing home to its climate) is to investigate your overall regional climate, the demands it makes, and the gifts it bestows on the houses in your locale.

Various regional climatic factors affect the heating and cooling of your house: the average ambient air temperatures, average precipitation (rain, snow, and hail), humidity (level of moisture contained in the air), wind speed and direction, and amount of sunlight available. All of these factors can add greatly to your heating and cooling problems if you ignore them when you design (or remodel) your home. Give them careful consideration, however, and you can design your home to use them selectively when they contribute to your comfort and screen them out when they don't.

For specific data on your regional climate, consult any or all of the following: your local U.S. Department of Commerce weather monitoring station, the highly technical ASHRAE (American Society of Heating, Refrigerating, and Air Conditioning Engineers) *Handbook of Fundamentals,* or the National Weather Records Center in Asheville, North Carolina.

For information on climate as it relates specifically to solar heating, turn to page 35, "Climate Considerations."

Microclimate. Once you are familiar with your regional climate, narrow your perspective and look closely at your **microclimate**—the weather patterns peculiar to your own site or house and its surroundings. A microclimate will probably share many of your region's climatic characteristics. But it will also have characteristics of its own, because of its individual topography, landscaping, and proximity to surrounding buildings, hills, lakes, or open plains.

To see how climate can vary within a small area, consider the city of San Francisco, whose general climate is often described as cool, temperate, and foggy. The Sunset district, fringing the ocean on San Francisco's west side, is indeed shrouded in fog much of the year. But the nearby Noe Valley district, nestled to the east beyond intervening hills, spends most of its days basking in bright sun.

Prevailing wind buffets house on bluff, creating cooler climate

Trees deflect wind from lower house; making climate warmer

Drawing 5: Microclimates: wind

Just as one district of a city may vary in climate from another, so an individual site can differ from its neighbor. While one house perches openly on a windy bluff, the house next door may shelter behind a stand of trees. The first house must be designed to fend off the wind, but the second will find the wind no problem. Manmade factors as well as natural ones can help determine a microclimate: if you live in the shadow of a

Shadow of apartment building creates cool microclimate for neighboring house

Drawing 6: Microclimates: sun

high-rise apartment, your house's climate will certainly differ from that of the house in the sun across the street!

Take time to observe the factors that determine your microclimate. As the seasons change, different aspects of your plot of land will come into focus. Some things to note would be: From what direction do the winter winds blow? Is your house or site protected from winds by structures or plants nearby? Where do summer breezes come from? Are they blocked from your house so they can't help cool it? Where are your trees and shrubs placed? Do they shade your windows or roof in summer, helping to keep the house cool? Do they block the sun's warm rays from the house in winter? Does a tall building or hill block your sunshine at certain times of the day or year? Morning or afternoon? Winter or summer?

Answering these and similar questions will tell you how best to take advantage of your climate and site in designing your home for natural comfort and preparing it for solar heat.

Facing the Sun

Of all the climatic elements you should consider in your home's design, the most important is the sun. Its relation to your house is especially crucial, of course, when you're planning on solar heat.

Your goals are to gain maximum advantage from the sun's warmth in winter, and to protect your home from too much summer heat gain. (In the hottest areas, of course, you may need to screen your house from the sun all year around.) To achieve these two goals, you must follow the sun's movements across the sky and orient your house accordingly.

The sun "moves" across our sky in two ways: along a daily east-west axis and along a yearly north-south path. It rises and sets farthest to the north on the summer solstice—June 21. That day, at noon, when the sun is at its highest elevation, it reaches its highest point of the whole year, appearing almost directly overhead. By September 21 (the autumnal equinox) the sun has completed half of its migration to the south; at noon on that day it is midway between its highest and lowest elevations.

By the winter solstice (December 21) the sun has reached its most southerly position and is at its lowest elevation that noon. At this point the sun hangs low above the horizon, appearing less than halfway up in the sky. March 21 (the vernal equinox) finds the sun back at its midpoint, on the way north again.

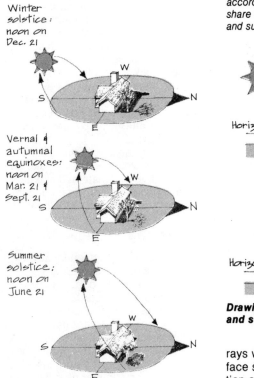

Drawing 7: Sun's annual path

At any point along the sun's annual route, its angle above your horizon is determined by your latitude—the farther north you live, the lower the sun will appear in the sky. In Santa Fe, New Mexico, the sun's noon altitude on the winter solstice will be at a 31° angle above the horizon, but in Boston, at noon on the same day it will be a mere 24° above the horizon.

As the winter sun rises in the southeast and pursues its low course across the sky to set in the southwest, the parts of your house that receive the greatest exposure to its

Drawing 8: Latitudes of U.S.
The farther north you live, the lower the winter sun is in the southern sky. The map breaks the U.S. into three general areas according to latitude. Homes in each area share the same basic sun angle, winter and summer.

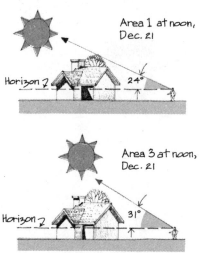

Drawing 9: Winter sun's angle: north and south

rays will be the walls and roof that face south. The greater the proportion of your total wall and roof area that is facing south, and the more directly south it faces, the more heat your house will derive from the winter sun.

In fact, as leading environmental architect Victor Olgyay pointed out, the house shape that gets optimum heating benefit from the sun, regardless of latitude or climate is "a form elongated along the east-west direction." If your house faces within 20° east or west of true south, it is suitable for solar heating, because it is within the range of optimum solar exposure.

Any walls that face north, on the other hand, receive no direct winter sunlight at all. And east and west walls receive direct sun only in the morning and afternoon, respectively. This leads to a fundamental principle for natural energy conservation and solar heating: *orient your house to face as near true south as possible.*

Rectangle on north-south axis gains least heat from sun

Square gains moderate heat from sun

Rectangle on east-west axis gains most heat from sun

Drawing 10: Facade orientation

Here we come up against a problem: Won't a house that faces south to benefit from the winter sun also suffer from overheating in the summer? Drawing 7 shows that in our northerly latitudes, the noon sun remains slightly to the south of us even in summer, threatening our south-facing walls with summer heat gains.

Except in the hottest and/or most humid parts of the country, this problem can be solved by adding shading devices that keep the sun from striking the walls in summer but permit it to do so in winter. (For more information on how this can be accomplished, see page 12.)

If you live in an area where summer heat is more of a problem than winter cold (as in Phoenix, Arizona, for example), you might want to make cooling rather than heating your first priority (see page 33).

Design Ideas for Climate Control

Once you've taken stock of climate and sun, you're ready to consider some energy-conserving design features to help make your house climate-responsive. Though not all of the following design ideas are involved directly in the process of solar heating, they make an important—sometimes invaluable—contribution in facilitating that process.

Before you plan any changes or additions to your home, be sure to consult an architect or contractor familiar with solar heating, or refer to our bibliography, page 93, for sources of more detailed information on energy conservation and solar heating.

Windows for warmth. Views, ventilation, and light are three benefits of windows. In a climate-adapted, sun-tempered house, there is a fourth: heat. Windows placed in a wall (or roof) that faces directly south become winter heat boosters.

On a clear winter day, they allow the low sun's rays to penetrate deeply into a room, warming the interior through a process known as the **"greenhouse effect."** The window glass admits the sun's short-wave radiation into the room, but when it strikes surfaces, they absorb it and, in turn, emit heat (long-wave radiation) that the glass will not transmit back outside. Instead the window traps the heat, enabling the room's interior to store it.

You will gain even more on heat savings if you concentrate *most* of your windows on the south wall, leaving the north wall relatively windowless (a few windows on the north wall can be useful for summer ventilation and for glimpses of a northern view, though they lose lots of heat in winter).

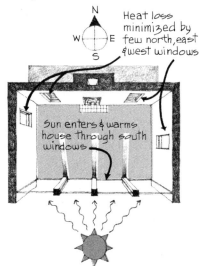

Heat loss minimized by few north, east & west windows

Sun enters & warms house through south windows

Drawing 11: Heat booster windows

East and west windows tend to lose more heat in winter than they gain from their brief daily exposure to the sun, so keep their numbers down or provide them with adequate movable insulation and summer sun-screens. A few east windows are useful, though, in that they admit the rising sun for early morning warmth. West windows have an added disadvantage: in summer they admit a lot of solar radiation during the hottest part of the day—the late afternoon. West windows should be avoided as much as possible without sacrificing view or ventilation, unless late-afternoon shading is provided.

If your best views are to the south, you're in luck. Your south windows will provide both heat and view. But if the views you want are to the north, east, or west, you must compromise, placing a few windows with care to feature the views and then facing the rest of the windows south for heat.

Fixed windows (ones that don't open) are good heat conservers, but be sure to allow enough openable windows or vents on all sides of your house to allow for cross-ventilation and summer cooling.

Zoned heating. Not all rooms in a house need to be exactly the same temperature; instead, you can adapt the heat level to the room's purpose, with quite a dividend in energy savings. A kitchen, for example, stays remarkably warm just from the heat given off by the refrigerator, stove, and other appliances; it may remain comfortable with less heat from a central heating system than a living room requires.

You can keep a bedroom cooler than a sitting room or study, because you only use the bedroom at night when you're snug under blankets. And in high-activity rooms such as studios and playrooms, the occupants generate heat on their own.

By considering which of your rooms need the most heat and which the least, you can divide the floor plan of your house into **temperature zones,** arranging warm rooms together in one part of the house, cooler rooms in another.

The warmest rooms should be clustered on the house's south side, where they receive natural heating from the sun. This zone might include the living room, dining room, and study. Cooler rooms could be grouped on the north side to form the second zone; here you might place the bedrooms, kitchen, and utility rooms. Bathrooms could be in the

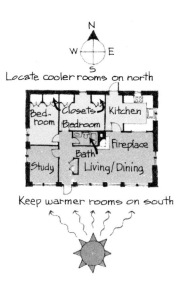

Locate cooler rooms on north

Bed-room | Closets | Kitchen
Bedroom
Bath | Fireplace
Study | Living/Dining

Keep warmer rooms on south

Drawing 12: Heat zoning

center, well protected from the elements by the surrounding house. By proper zoning, you can close off the heat to some rooms for long periods—bedrooms by day, other rooms by night. (Doors between zones must close tightly.)

Temperature zones may be treated in either of two basic ways. Some climate-adapted houses isolate the zones from one another, completely controlling the temperature of each. These houses even have insulated walls and doors between zones, in order to seal the heat into the warmer section.

Other houses—especially those heated by direct sunlight—have virtually no walls or barriers between rooms at all. They permit the free circulation of heat from zone to zone, relying on the arrangement of the rooms to moderate the temperature as needed. The bedrooms in these

Heat moves easily through open loft

Minimal interior walls allow heat to circulate freely throughout house

Drawing 13: Open-plan house for easy heating

houses are often placed in an upper story or loft, so that the downstairs heat simply filters up to them.

Earth berms and underground houses. Berming your house means banking some of its walls with earth, or even partially burying it in the ground. The earth mass protects the house from the wind and stabilizes the indoor temperature by moderating the swings of winter and summer outdoor temperatures.

Earth is piled against house to keep heat inside; landscaping holds earth in place

Drawing 14: Berm styles: mounded

Solar-oriented houses often have high berms of earth mounded against their north walls, and lower berms pressing their east and west walls. Sometimes a solar house is actually built into a south-facing hillside, so that the hillside itself becomes the buffer. Other solar houses go beyond berming—they have sod roofs or are built almost entirely underground.

Piled earth is held against house by retaining wall; berm helps keep house warm

Drawing 15: Berm styles: wall-supported

The earthen envelope surrounding the house gives a year-round stable temperature of about 55° to 60° F. (the constant temperature of the

House is dug into south-facing hillside for stable temperatures

Drawing 16: Berm styles: hillside

earth below the frost line). It has the added benefit of cutting down on noise intrusion and providing extra gardening space on limited lots.

Berms and underground construction are particularly useful in solar design, as they lessen the heating burden placed on the solar system, and can even be used to "design" a more advantageous microclimate by blocking wind and encouraging sun (see Drawing 17).

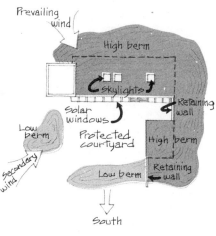

Drawing 17: Climate control with berms

Underground building is not limited to rural areas; it is perfectly possible on a suburban lot, though you may need to import extra earth from somewhere else (it takes a lot of earth to cover a house!).

Also, contrary to popular impression, underground houses needn't be tomblike: well-placed skylights, windows, and vents can keep their interiors light and airy. The south sides of many of these houses are left open to the sun for heating.

To determine the height and mass of your berms, you'll need to balance your need for protection and your desire for windows. You might want to pile a berm almost to the roofline

on the north side, while on the east and west, you'd want to taper the berms down to waist height to allow windows for views and light.

If you make use of earth berms or underground construction, several structural precautions are necessary to prevent damage to your house.

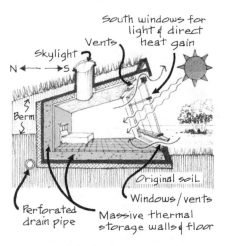

Drawing 18: Underground house
Berm completely covers north wall and roof; only skylights protrude, for light, ventilation. Additional vents in south windows, roof peak keep air fresh, prevent heat buildup. Drain pipes, buried in berm, carry moisture away from house to avoid damage, leakage.

First, your walls (and, where applicable, your roof) must be designed to bear the earth's weight (which is considerable, especially when rain soaks the ground). The walls must also be insulated on the *outside*, between wall and earth down 2 to 4 feet (well below the frost line), so that heat from the house's interior will not creep out to warm the surrounding earth.

Next, special waterproofing and drainage systems must be installed before the earth is packed into place so that it won't soak up and retain rainwater, holding it against your walls and risking structural damage and leakage. (A high water table prohibits underground construction.) As a final step in the project, the earthworks should be landscaped, not only for appearance but also to hold the earth in place. Before embarking on a berming project, you should consult a structural engineer or architect familiar with berms and underground construction (see our bibliography, page 93, for sources of information).

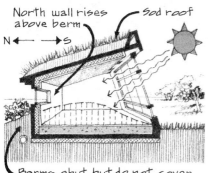

North wall rises above berm

Sod roof

N ← → S

Berms abut but do not cover house, giving more light & ventilation but slightly less temperature stability

Drawing 19: House with sod roof

Sod roof caps a high-bermed house; structure rises out of ground at shoulder height, has small north windows at eye level in addition to heat-gathering south windows.

Movable insulation. The great heat thieves in a well-insulated house are the windows, even when they're double or triple glazed—made with two or three layers of glass with a dead air space about 3/16 inch wide between layers for insulation.

Windows become a special nighttime and cloudy-weather problem in climate-adapted and solar houses that have large, south-facing window areas. If you're relying for winter heat on the unreliable sun, just the overnight heat loss through your windows can have serious consequences, draining your heat resources when there is no certainty of quick replenishment.

One solution to this problem is **movable insulation** for the windows: insulation that covers the windows when needed but is removed when you want light or heat gain. Types of movable insulation are as various and inventive as window styles themselves. Solar designers often create insulation as part of a solar heating

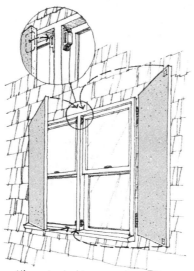

Hinged shutters made of insulation swing open or shut

Drawing 20: Insulating shutters

system, but many homeowners design and construct their own to suit their needs and windows.

The basic principle is usually the same: a panel or curtain made of some form of insulative material, such as fiberglass, covered styrofoam or polystyrene, or several layers of heavy cloth. When in place, the panel or curtain fits flush against the window pane and snugly within (or over) the frame, leaving no chinks around its edges through which heat can pass.

Panel slides into slot in wall during day; pulls out along track to cover window at night

Drawing 21: Sliding insulation panel

Some types of insulating panels are attached to a window's exterior; others are mounted inside. (Flammability makes styrofoam and polystyrene hazardous as interior insulation.) Some panels swing open like shutters; some fold aside accordion-style; and some slide away from windows on tracks, like Japanese shoji screens. Others are removed from the windows entirely during a sunny day and put up again only when needed at night or during a prolonged storm.

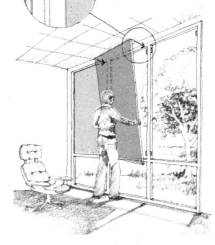

When panel is in place, small cupboard-door magnets hold it to window frame; when not needed, panel is removed, stored in cabinet

Drawing 22: Removable insulation panels

One variety of hinged, foil-covered insulating panel does double duty: closing against the window to insulate at night, and swinging up (or down) by day to bounce extra sun into the house off its reflective inner surface (see Drawing 23, page 12).

Many insulating devices are manually operated. Though they tend to be fairly simple in design and they cost nothing to operate, they require your time and your presence (at nightfall or when a storm blows up) to put them in place.

(Continued on next page)

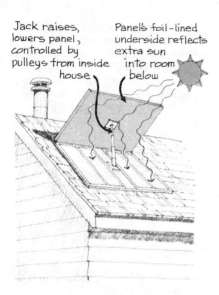

Jack raises, lowers panel, controlled by pulleys from inside house

Panel's foil-lined underside reflects extra sun into room below

Drawing 23: Reflective skylight insulation

Other devices move themselves. Some are temperature activated and operate automatically. Others are mechanically operated and controlled by thermostats. When the thermostat senses a certain rise in temperature, it opens the panels. It closes them again only when the temperature drops beyond a specified point.

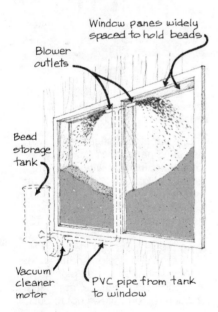

Window panes widely spaced to hold beads

Blower outlets

Bead storage tank

Vacuum cleaner motor

PVC pipe from tank to window

Drawing 24: Blown styrofoam insulation

The window insulates itself: vacuum cleaner motor blows tiny styrofoam beads from storage tank to fill compactly the space between window's two panes. When sun shines, motor reverses, sucking beads back into tank.

These devices involve some operating costs (as well as initial cost), but they save time and operate even in your absence.

One extraordinary automatic insulating device was designed by the solar design firm Zomeworks as an integral part of the window itself. The "Beadwall" uses a vacuum cleaner motor to blow tiny styrene beads into the wide space between the window's two panes of glass, filling the space entirely to form the insulating barrier. When the sun begins to shine, the motor sucks the beads back into storage drums until they're needed again

Another ingenious automatic Zomeworks device, "Skylid," is especially good for out-of-reach skylights. It uses Freon in small canisters to activate insulated

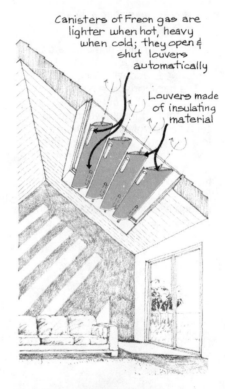

Canisters of Freon gas are lighter when hot, heavy when cold; they open & shut louvers automatically

Louvers made of insulating material

Drawing 25: Insulating louvers

Louvers of insulation for out-of-reach skylights revolve open when sun is out, close to cover skylight when sun is gone. They're controlled by canisters of heat-sensitive Freon gas attached to opposite sides of each louver.

louvers mounted inside the house against the window. The louvers open and close automatically as the sun comes and goes.

All types of movable insulation are useful to maintain summer coolness as well as winter heat. In the summer,

you simply reverse the insulating schedule, leaving the panels over any windows exposed to the daytime sun, and removing them at night. Used this way, movable insulation helps to minimize unwanted heat gain, especially through skylights and south and west windows.

Selective sunscreens. A house with lots of windows on its south, east, and west sides will heat up like an oven in the summer unless the windows are shaded. Insulating panels (see "Movable insulation," preceding) will do the job, but at the cost of view and ventilation—a high price to pay on a lovely summer day.

Highly reflective interior blinds or curtains will help to screen some sun out, but they also block the view. What is worse, the sun still enters the house between curtain and window, adding appreciable heat despite the appearance of shading.

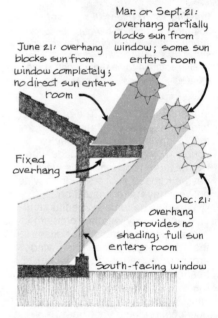

Mar. or Sept. 21: overhang partially blocks sun from window; some sun enters room

June 21: overhang blocks sun from window completely; no direct sun enters room

Fixed overhang

Dec. 21: overhang provides no shading; full sun enters room

South-facing window

Drawing 26: Fixed overhang shading

For south windows, exterior shades and overhangs block the summer sun much more effectively than interior shades because they keep the sunlight from reaching the windows at all. Many solar houses use permanent overhangs to provide summer shading: extended rooflines or wooden awnings or trellises that project out several feet over their south windows. These overhangs are sized according to seasonal sun angles, so that they cast complete shade over the windows from late May to early September but allow the sun to enter the windows throughout the winter months. **Fixed overhangs** have one drawback, however: sun angles do not correspond neatly with high temperatures or the need for shading. The sun is highest on June 21, but summer weather is usually hottest in August and early September, when the sun is already lower in the sky. An overhang long enough to screen out this late-summer sun will also block the sun in late March and early April, when it is still welcome to take the edge off the spring chill.

Some solar designers compromise with a narrower overhang; they live with a little less spring sun and a little more late-summer sun than might be ideal. Others install **movable shading devices,** which can be adjusted according to the sun's angle and the shading need.

The simplest example of a movable shading device is a canvas awning that retracts to admit April sun and then extends to provide deep shade in August and September. In October's first chilly weather, it can be folded against the wall or removed from its frame for winter, leaving the windows open to the sun.

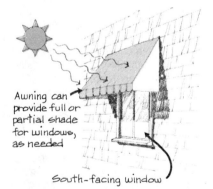

Awning can provide full or partial shade for windows, as needed

South-facing window

Drawing 27: Movable shading: awning

Those who find awnings unappealing have other options. For example, a wooden overhang can be designed with a hinged flap that folds back to admit April sun, and extends out to

Hinged overhang has two positions: extended for full Sept. shade, or folded back for March sun

South-facing window

Drawing 28: Movable shading: hinged overhang

full width for August and September shading. Or a narrow fixed overhang can have canvas, bamboo, or woven grass blinds or roller shades mounted at its edges; these can be pulled down for shading and rolled back up to admit the sun.

Roll-up blind is adjustable for morning or afternoon shading of east & west windows, or fall shading of south windows

East, west, or south-facing window

Drawing 29: Movable shading: blinds

Still another ingenious idea is to build an open trellis that projects out above the windows and grow deciduous vines such as grape, morning glory, or Virginia creeper on it. When the vines leaf out in spring, they begin to shade your windows; in midsummer they provide dense shade; and by early fall they're ready to drop their leaves and let the sun shine into the house once again.

East and west windows pose a different shading problem from south windows, because the morning sun is low in the eastern sky and the evening sun is low in the west every day, winter and summer. As a result, overhangs are of very little use in blocking the sun's rays. Overhangs equipped with shades or blinds that

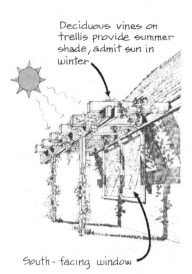

Deciduous vines on trellis provide summer shade, admit sun in winter

South-facing window

Drawing 30: Movable shading: vines

can be raised or pulled down as needed are one solution. Another is to mount trellises vertically in front of the windows and grow vines on them.

For more detailed sources on design and sizing of shading devices, see our bibliography, page 93.

Landscaping for heat control. The same kind of thought that goes into adapting a house to its climate should also be expended on the garden, because landscaping can do a lot to help control a house's heat loss or gain.

Take the problem of summer heat gain, for example. If your house is surrounded by a barren field of asphalt—or even bordered by an unshaded drive or patio on its south or west side—it's bound to simmer in the summer sun, because the concrete will soak up much more heat than plants and lawns ever would. Patios and walkways should be shaded by trees, awnings, or roofing wherever summer heat is a problem.

Thoughtful landscaping can help alleviate shading problems for your east, south, and west windows. Under ''Selective sunscreens'' (preceding), we discussed use of shade trellises for this purpose. Deciduous trees also make excellent selective sunscreens, because they leaf out and cast a dense shade precisely when you need it most: in the summertime. In winter, they shed their leaves obligingly and present a minimal obstacle to the desirable winter sun. The deciduous trees should be planted on the east and especially the west side of the house, where they will contribute the most summer shade to the appropriate windows. (Evergreens should not be used as sunscreens because they continue to block the sun from the windows in the winter.)

Wherever wind is a problem, landscaping again offers a solution—trees and hedges can be planted to block its assault on your house. For example, dense plantings of evergreen coniferous trees to the north and northwest of your house can make effective windbreaks, especially if they are tall and planted so that they deflect the flow of air over and around the house.

Winter heating

Prevailing winter winds blocked by evergreens

Deciduous trees lose leaves, allow low sun to reach house windows

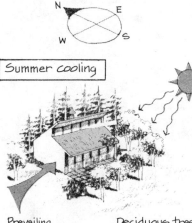

Summer cooling

Prevailing summer breezes provide natural cooling

Deciduous trees positioned to shade patio, house.

Drawing 31: Landscaping for heat control

Even mounds of earth, strategically shaped and landscaped, can help steer errant winter winds away from your susceptible windows and walls.

The Basic Systems

If you've met the sun halfway with a weather-tight, climate-responsive house, you're ready for the final step: solar heating. The two main categories of solar heating are **passive** (architectural) systems and **active** (mechanical) systems. Some designers combine features of both into a third approach called a **hybrid** system, in which fans, ducts, and/or extra storage area facilitate the heating function of a passively designed house. Into these categories fall a number of different system types, each with its own architectural requirements, esthetic choices, operating techniques, and influences on your life style. Some system types can only be accommodated in a new house designed specifically for solar heating. Others can be retrofitted (added to an existing house built originally for conventional heating).

You can live in a solar-heated home that is completely conventional in appearance and functional except for panels on the roof and a tank or rock bin in the basement. You can also follow a life style centered on the sun's daily and yearly rhythms, in a house whose primary design influence is its solar heating system. The choice is yours. To help you make it, we will describe the basic system types, how they function, and how they may affect your life style and your house's style.

Passive: The House as Heater

To be passive means not just to be inactive, but to be receptive to an external force—in our case, the sun. In the passive approach to solar heating, the heat flow takes place by natural means: conduction, radiation, and convection. The house is designed to allow the sun to warm its interior directly, without intermediating machinery.

The house also stores heat within its own structure for use at night and on cloudy days, and distributes the heat with a minimum of mechanical aid. In short, the house becomes its own heating system. This puts

passive solar heating into the realm more of architecture than of mechanical engineering, since the design of the house creates the heater.

The three functions of any solar heating system are **collection** of the sun's energy, **storage** of the energy as heat, and **distribution** of the heat throughout the house when needed.

In a passive system, collection of heat is the easy part. The aim is to maximize solar heating. Some passive systems are designed for **direct gain** of the sun's heat: extending the house along an east-west axis, filling most of the south wall with windows for maximum solar exposure, and encouraging the sunlight to penetrate through those windows into the house, making use of the "greenhouse effect" to trap heat. North windows are kept to a minimum; east and west windows are used with discrimination (see "Facing the Sun," page 7, and "Windows for warmth," page 9). Other passive systems collect and store the sun's heat in one part of the house and utilize natural heat movement to warm the rest of the house. These are known as **indirect-gain** systems.

The main challenge of direct-gain heat collection is to get direct sunlight into as many areas of the house as possible, so that heating will be even. (In an indirect-gain system, the object is to admit sun to strike the crucial storage areas, using some of the same techniques as a direct-gain system.) There are various ways to do this: build a shallow house on a long east-west axis; stack a shallow house vertically to face south; give rear rooms south-facing clerestory windows; step a house up a south-facing hillside with windows on each level to draw light down into the rooms; use skylights in a south-facing roof to bring sun deeper into the rooms, to heat interior walls; or devise a fan system to pull hot air from sunlit south-facing rooms to northern ones with no sun exposure.

Storage and distribution present more problems than collection for passively heated houses, because the storage must be contained in the structure of the house itself, and the distribution must occur naturally. The solution is called **storage mass**: You build a house using dense materials

Long east-west axis gives good southern exposure

Tall house maximizes vertical southern exposure

Sun reaches all levels of house stepped up south-facing hillside

South-facing skylights admit sun to rear of house

Drawing 32: Passive heat collection

with high heat-holding capacity—masonry, adobe, concrete, stone, and water—for the walls and floor (or, in the case of indirect systems, in the area specified for storage). The mass absorbs heat from the sunlight entering the windows, storing it until the sun sets or goes behind a cloud. Then, as the room air cools, the mass radiates its stored heat back out into the rooms (see Drawing 35, page 16).

The amount of mass needed in a passive house depends on the amount of sunlight available, the amount of heat needed to keep the house comfortable in its climate, and the materials used to build the mass. (Different materials have different heat storage capacities: concrete can store 30 Btus of heat—see glossary—per cubic foot per 1°F. rise in temperature, whereas brick holds 25 Btus and water holds 62.4.) If the house is properly designed, its mass should be able to store heat adequate for at least 1 night's needs, and maybe up to 1 or 2 cloudy days' heat. Even this much storage is rarely enough to provide all of a house's heating needs; a backup heater is almost always necessary for prolonged periods of bad weather.

A passive house should rarely overheat on a winter's day, as the storage mass will absorb most of the excess heat.

"Mass" in a passive house often means walls and floors up to 2 feet thick (sometimes with water-filled steel drums or polyethylene bags inside them to increase the quantity of heat they can hold).

These massive structural elements are insulated on the outside, just under the exterior layer of stucco or paneling, around the perimeter of the floor slab or foundation, and (in very cold climates) beneath the slab itself. Since it is *outside* the mass, the insulation does not block the incoming southern sunshine from heating the storage mass, but it helps to retain the heat in the mass.

The **R-values** (resistance to heat flow) of insulations used for solar houses range from absolute minimum R-11 up to R-19 for walls and floors, and from minimum R-19 up to R-30 for ceiling or roof. A common type of exterior insulation is the new generation of efficient, high-density rigid foam panels.

Often a passive house in a cold climate requires some form of movable insulation for its windows, in

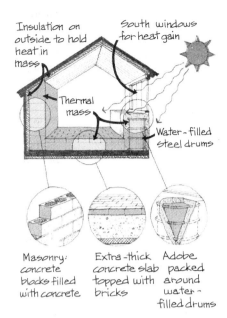

Masonry: concrete blocks filled with concrete

Extra-thick concrete slab topped with bricks

Adobe packed around water-filled drums

Drawing 33: Thermal mass for heat storage

addition to the other insulation, in order to hold onto all its low-intensity solar heat. For ideas on this point, see "Movable insulation," page 11.

Mechanical forms of heat distribution, such as small fans and ductwork, are often used to help move heat to cooler parts of a passive house, making the system technically a "hybrid" of active and passive techniques. Many passive solar designers prefer to rely on natural heat movement: radiation directly from the storage mass, conduction through interior connecting walls, and convection from room to adjoining room, assisted by vents and dampers to control the flow of air.

Passive heating systems operate at low temperatures compared to those produced by a conventional furnace or an active solar heating system (see page 21). One reason is that the sun's heat is of low intensity when collected; another is that the storage walls and floor do not build up heat to high temperatures over a period of days, since whatever heat they store in the day is generally released at night. As a result, these systems perform best in climates such as those in New Mexico or Colorado, where the coldest weather is often the brightest, and the sun can be counted on to replenish the supply of heat regularly. (Passive cooling is also most effective in these areas, where dry air and cool summer nights are the norm.)

Nevertheless, if the house is a good heat-conserver, passive heating is feasible in virtually any part of the country. Its low operating temperatures actually give passive heating a certain advantage for such hazy areas as much of New England, because a passively heated house can collect some heat even when the sky is hazy or cloudy and the sun's radiation is diffuse. Also, heating a house with a large surface at a low temperature is more comfortable, and less heat is lost from a low-temperature collection and storage mass, making the passive system very efficient.

Because the heating system in a passive solar house is actually a major structural element, most passive solar houses are designed and constructed specifically for solar heating. It would be quite difficult to **retrofit** (remodel) a conventionally heated house for most passive solar heating systems unless major changes or additions were planned.

The biggest expense in a house built for passive solar heating is the literally *massive* construction of its walls and floors. Another expense not to be ignored is the auxiliary conventional heating system required by law in most areas of the country, and usually necessary to supplement the passive solar system in periods of extremely bad weather (see "The Backup: Auxiliary Heating," page 31).

Though a house constructed this way may cost more initially than a conventional thin-walled house, its generally smaller size helps keep costs down. And, once it is built, its savings in heating (and cooling) bills can eventually offset the added initial cost. Unlike active solar heating systems (see page 21), most passive systems involve very little expensive hardware in addition to the house itself, so by the standards of solar heating in general, their costs are rather low. Maintenance costs are also lower.

Living in a passively heated home requires a certain adaptation to a life style most of us are unused to. The temperature in a passive house fluc-

Insulating shutters help hold in heat

Woodstoves as backup heat

Sweaters play a part in passive solar heating

Drawing 34: Adapting to a passive solar house

tuates much more than that in a conventional home. You may expect daily swings of 5° to 15°F. in winter—a room may be 75° after a day's heat collection, but by the next morning it could be down to 60°. That means you get used to throwing on a sweater (and peeling it off again), lighting wood fires for extra warmth, or even living with a little discomfort occasionally.

Your time and attention are often required for various daily routines: opening and closing vents, dampers, insulating panels, and doors according to the sun's position, the weather, and your need for heat.

As mentioned in "Zoned heating," page 9, many passive homes have a very open floor plan, with few floor-to-ceiling walls between rooms, and with semiprivate sleeping lofts instead of enclosed bedrooms. Different floor levels and low partitions are used to define rooms, leaving the air space unobstructed so heat can travel freely throughout the interior of the house.

Passive homes also tend to be on the small side, to keep the heat demand down. For some people this small space and lack of privacy requires an adjustment, but many owners of passive houses say that their open homes encourage a warm feeling of community. They enjoy their sun-centered life style and find "operating" their hard-working homes not at all arduous, but on the contrary—a pleasure.

Though similar in their basic approach to solar heating, passive

system types vary somewhat in their means of collecting heat, the design of their storage, and in their distribution systems. We will describe the main variations briefly. (For sources of more information on the design and operation of passive systems, consult our bibliography, page 93.)

Direct-gain systems. The simplest kind of passive house uses the direct-gain approach: it has a large window area facing south for direct solar penetration into the living

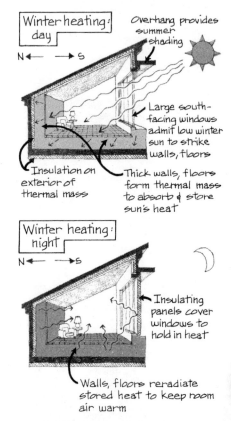

Winter heating: day

N ← → S

Overhang provides summer shading

Large south-facing windows admit low winter sun to strike walls, floors

Insulation on exterior of thermal mass

Thick walls, floors form thermal mass to absorb & store sun's heat

Winter heating: night

N ← → S

Insulating panels cover windows to hold in heat

Walls, floors reradiate stored heat to keep room air warm

Drawing 35: Direct-gain heating

space, with an overhang for summer shading and window insulation for winter nights (see Drawing 35). The thick floors and walls are built of concrete, stone, masonry, or adobe, for heat storage. The floor and walls may be inlaid with dark tile or stone, painted a dark color, or left plain.

The north wall may have no windows at all, or just enough to provide cross-ventilation for summer cooling in combination with the south windows. East windows may be included for winter morning heat; west windows are used generally to frame a great view. (Shading must be available for summer use.)

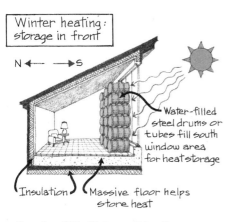

Drawing 36A: Water wall heating
Ranked just inside window, water-filled steel drums pile atop one another, round ends facing sun. Drums gather, store heat during day, reradiate it into room at night. (A wall of vertical tubes can do the same.) Insulating panels for windows block nighttime heat loss.

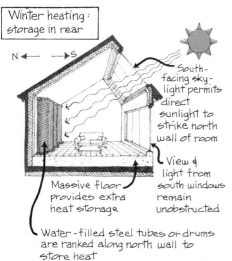

Drawing 36B: Water wall heating
To admit south view or light, water-filled steel tubes (or piled drums) are placed against rear north wall, absorbing and storing heat from sunlight striking through south skylights. Night heat is held in by skylight and window insulation.

Because a direct-gain system admits so much sunlight into the house, glare and fading of furniture may be problems.

Water walls. In some passively heated houses, water rather than stone or concrete is the material used for heat storage mass. The water is usually contained in drums, tanks, or columnlike tubes (made of steel ducting or highway culvert, fiberglass, or concrete lined with polyethylene) with dark, heat-absorptive surfaces, placed inside the house where they catch sunlight entering the south windows (see Drawings 36A and 36B). Another water wall system sandwiches the water, inside a polyethylene liner, between two thin "walls" of preformed concrete erected right inside the south windows. This system works like a Trombe wall (see below) but is much more efficient.

The main advantage of water over the other storage materials is its efficiency: it picks up heat faster, holds more of it, and releases it more readily, requiring considerably less storage space.

Using water as the heat storage medium, however, creates a risk of leakage if the container is metal. The danger can be minimized by adding anticorrosive stabilizers in the water, or by lining the containers with polyethylene.

With insulating panels and exterior shades or screens for skylights and south windows, a water "wall" can become an effective summer cooling system in areas where days are hot

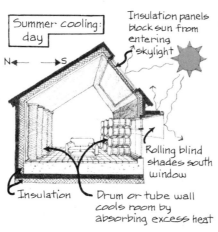

Drawing 37A: Water wall cooling: day
Keeping daytime cool, house is shaded to prevent sun from heating water-filled drums or tubes. Meanwhile, drums or tubes cool room air by absorbing any infiltration of outdoor heat.

but nights are cool, and humidity isn't a problem. With proper shading and venting, the cooled water in a water wall (see Drawings 37A and 37B) can keep the interior temperature of a house down to 75° to 80°F., even when it's 115° outside.

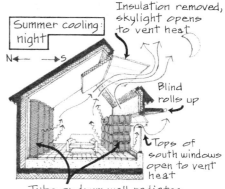

Drawing 37B: Water wall cooling: night
House opens up to cool at night: any buildup of heat in drums or tubes is radiated to room, rises, and escapes out open skylights, windows.

Trombe walls. Named for Dr. Felix Trombe, one of its developers, this passive heating system features a concrete, stone, or masonry south heat-storage wall up to 16 inches thick. It may or may not have glazed window openings in it, but it usually has vents at regular intervals both along the floor and just below the ceiling of each room. (A Trombe wall without vents can be used to store heat for nighttime use and to stabilize interior temperatures, but it won't provide much daytime heat.)

The entire exterior face of this wall is painted black or some other dark, heat-absorbing color; it is also covered by one or two layers of glass, mounted several inches out from the concrete wall.

The Trombe wall with vents performs two heating functions in winter—a daytime convective heating loop and nighttime radiation of the

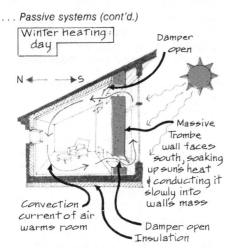

Winter heating: day

Damper open

Massive Trombe wall faces south, soaking up sun's heat & conducting it slowly into wall's mass

Convection current of air warms room

Damper open

Insulation

Drawing 38A: Trombe wall heating: day

Convection current starts when sun heats air between Trombe wall and outer glass. Hot air rises, enters room through vents at wall's top, gives off heat to room. Cooler air from floor is pulled through Trombe's bottom vents, recommences heating-and-rising cycle.

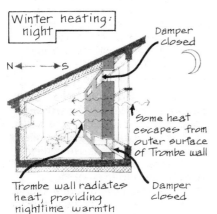

Winter heating: night

Damper closed

Some heat escapes from outer surface of Trombe wall

Trombe wall radiates heat, providing nighttime warmth

Damper closed

Drawing 38B: Trombe wall heating: night

Heat gathered in Trombe wall penetrates to wall's north side by late afternoon; it radiates gradually into room during night, providing warmth. Dampers at top and bottom vents in Trombe are closed to prevent convection current (see above) from reversing to cool room.

heat stored in the wall (see Drawings 38A and 38B). The vents in the Trombe wall that complete the convection loop must be fitted with dampers so that at night, when the room air is warmer than the outside air, they can be closed. Otherwise, the convection current would reverse itself, drawing warm room air out through the top vents, cooling it next to the glass wall, and sending it, cold, back into the room through the lower vents. (Also see Clearview collector on page 68.)

The Trombe wall has a summer cooling function as well as its winter heating ones. During the day, the upward convection between the glass and the concrete wall draws heat out of the house and brings cool air in (see Drawing 39). At night, all vents are opened to encourage the convection loop to reverse, cooling the house air between Trombe wall and glass.

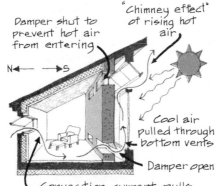

Damper shut to prevent hot air from entering

"Chimney effect" of rising hot air

Cool air pulled through bottom vents

Damper open

Convection current pulls cool air through open north windows & across room

Drawing 39: Trombe wall cooling

Cooling takes place when sun-heated air rises between Trombe wall and south window, escaping outdoors through opened vents at window's top. This air pulls room air after it through bottom vents in Trombe, in turn pulling cool air from shady side of house, through north windows, and across room.

Greenhouses. A solar greenhouse starts out just like any other attached greenhouse: a glass, fiberglass, or polyethylene-covered structure attached to the side of a house, in which you can grow plants that wouldn't survive the climate outdoors. There the resemblance ends. Unlike other greenhouses, a solar greenhouse derives all its heat from the sun. And it can be designed to share its heat with the house to which it is attached.

Solar greenhouses are usually attached to a south-facing wall of the house, so they catch the winter sun. They can also be located on east or west walls as long as they have some southern exposure. They generally have massive stone, concrete, or masonry floors, and sometimes include black-painted water-filled drums ranked against the house wall or lined up under the south windows of the greenhouse. Sometimes the wall between the greenhouse and the house is built massively of adobe, concrete, or other masonry. All these elements act as heat storage and prevent overheating; the connecting wall can even be used as a Trombe wall (see preceding). In addition, the perimeter of the greenhouse is insulated on the exterior to a depth of 3 to 4 feet. In severely cold climates, the entire greenhouse is often underlaid with rigid insulation beneath the slab and the built-in earth planting beds, to prevent loss of heat to the surrounding ground and to transform the whole slab and earth beds into more heat storage.

Unless the wall between house and greenhouse is built for use as a Trombe wall, it must be well insulated on the side that faces into the house itself, so that the house doesn't lose heat to the greenhouse at night, yet the wall can be used to store heat for greenhouse use. A door, windows, and sometimes vents connect house and greenhouse. These are opened for heat gain and closed when necessary to prevent heat loss to the greenhouse (see Drawing 40).

On some of the coldest winter nights in severe climates, though, a greenhouse's own storage mass may not be enough to keep it above freezing without insulating shutters or auxiliary heat.

A solar greenhouse can assist in summer cooling as well as winter heating, by encouraging a convection current to draw heat out of the house and pull cooler air in from the north (see Drawing 41).

Some greenhouses are covered in winter with polyethylene film instead of glass to create the necessary warm climate. In the summer, the

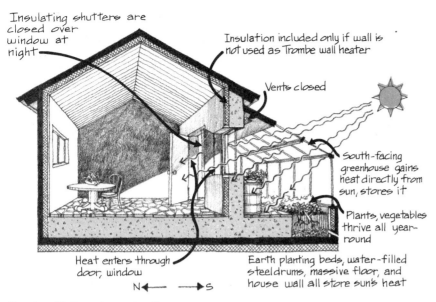

Insulating shutters are closed over window at night

Insulation included only if wall is not used as Trombe wall heater

Vents closed

South-facing greenhouse gains heat directly from sun, stores it

Plants, vegetables thrive all year-round

Heat enters through door, window

Earth planting beds, water-filled steel drums, massive floor, and house wall all store sun's heat

N ←— —→ S

Drawing 40: Greenhouse heating

Attached solar greenhouse is wonder worker: it heats itself by gathering and storing sun's heat, reradiating it at night; it also adds heat to main house by conduction through wall and convection through doors, windows. If greenhouse's temperature swing is great, main house may be insulated from greenhouse, closed off at night to prevent heat loss to gradually cooling greenhouse. If greenhouse stores enough heat, or if wall between house and greenhouse acts as Trombe wall (see page 17), house need not be shut off, but may gather additional nighttime heat from greenhouse.

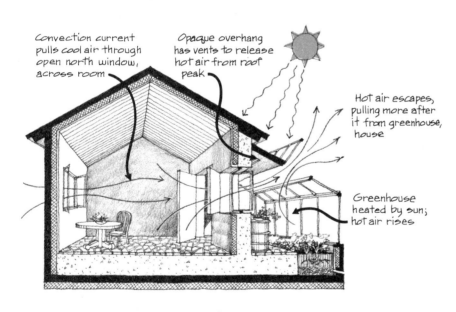

Convection current pulls cool air through open north window, across room

Opaque overhang has vents to release hot air from roof peak

Hot air escapes, pulling more after it from greenhouse, house

Greenhouse heated by sun; hot air rises

N ←— —→ S

Drawing 41: Greenhouse cooling

To avoid summer overheating, greenhouse's vents and windows, high on east and west sides and across front of overhang, are opened. Sun's heat creates "chimney effect": hot air rises out of greenhouse, pulling cooler air by convection in through house's north windows, across room, and through open doors and windows into greenhouse. Opaque overhang shades house to prevent heat buildup in wall (especially important if wall is not insulated from house).

polyethylene is taken down and stored, creating a pleasant patio full of plants where the greenhouse used to be.

In some designs a greenhouse almost "takes over" the house, becoming so large that it could almost rank as a "direct-gain" system, were it not separated from the rest of the house by walls to control the heat flow. Such a greenhouse is designed to provide the major portion of a house's heat.

Most solar greenhouses have several purposes in addition to passive solar heating: They can produce food crops almost year-round, even in severely cold climates; they can be an esthetic, gardenlike addition to the house; they serve as a convenient airlock to prevent undue heat loss when the house door is opened; and in winter the plants they contain help humidify and purify stale, dry house air.

Because a solar greenhouse is usually a self-contained structure attached to a house, it is one of the easiest to retrofit (adapt to an existing house) of all the various passive heating systems, provided the house has a wall with a southern exposure to attach it to.

Roof ponds. Not really ponds at all, but rather open troughs or polyvinyl chloride (PVC) bags filled with water and resting on black underliners, roof ponds are another passive heating system that uses water for heat storage. Set into the roof atop a sturdy ceiling, they capture and store solar heat, radiating the heat directly through the ceiling to the rooms below. The rooms themselves are often built with extra mass to absorb and help store that heat (see Drawing 42 on page 20).

Some roof ponds, such as those designed by Harold Hay, who originated the roof pond idea, have insulating covers that roll out on tracks at night, then roll away by day to admit the sun. Others have lids of movable insulation that serve an extra purpose as well. Their inner surfaces are lined with reflective foil; during the day, when the lids are lifted, the foil bounces extra sun down onto the ponds, increasing their heat intake appreciably.

(Continued on next page)

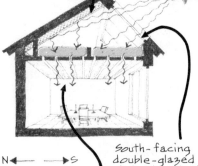

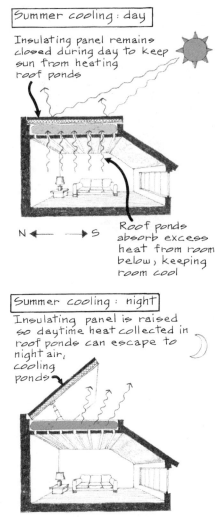

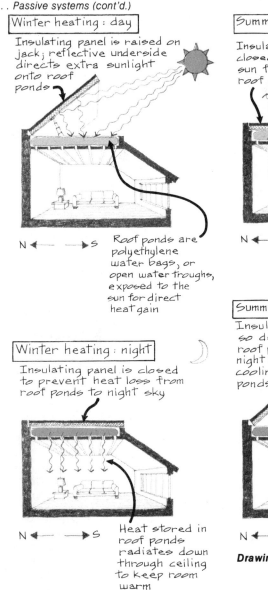

Winter heating: day

Insulating panel is raised on jack; reflective underside directs extra sunlight onto roof ponds

N ◄ — ► S

Roof ponds are polyethylene water bags, or open water troughs, exposed to the sun for direct heat gain

Winter heating: night

Insulating panel is closed to prevent heat loss from roof ponds to night sky

N ◄ — ► S

Heat stored in roof ponds radiates down through ceiling to keep room warm

Drawing 42: Roof pond heating

Summer cooling: day

Insulating panel remains closed during day to keep sun from heating roof ponds

N ◄ — ► S

Roof ponds absorb excess heat from room below, keeping room cool

Summer cooling: night

Insulating panel is raised so daytime heat collected in roof ponds can escape to night air, cooling ponds

N ◄ — ► S

Drawing 43: Roof pond cooling

Foil-lined backing of north roof's insulation reflects extra sunlight down onto roof ponds

N ◄ — ► S

Heat radiates from roof ponds to room below

South-facing double-glazed skylights admit sun to strike roof ponds, help prevent freezing

Drawing 44: Roof ponds for freezing climate

This is an important aid, because the horizontal pond surfaces have a fairly limited exposure to the low winter sun (much less than would a vertical surface), and the near-vertical reflectors intercept lots of useful sunlight.

Roof ponds have a cooling action that works especially well in climates with cool summer nights and low humidity. The insulating lids are opened at night to release any heat accumulated in the water. (Open troughs have the advantage of allowing evaporative cooling to take place, as the water is exposed to the air.) During the day the lids are closed to allow the cooled water to absorb heat rising from the house below (see Drawing 43).

In cold climates, the roof ponds need more protection than just insulating panels to prevent freezing problems. The ponds cannot remain exposed directly to the air. One solution using water bags is to build a slanted Cape Cod-style roof over the bags (see Drawing 44), with large skylights in the roof's south side to admit the sun, and reflectors lining the north roof to beam extra sun down onto the ponds. (This arrangement improves heat collection in northern latitudes, where the sun's angle is lower and reflectors are more necessary.) On winter nights and summer days, the skylights must be covered with some form of movable insulation to prevent heat loss or gain.

Roof ponds are even heavier than a sod roof; they require extra structural support.

Roof ponds have some drawbacks, too: the ultraviolet rays in sunlight eventually break down PVC bags, which must be replaced periodically or they may leak. Also, the ponds can heat only the rooms that are directly below them, if no provision is made for channeling heat elsewhere (as to the lower floor of a two-story house).

Thermosiphoning air collectors.
Thermosiphoning is another way of saying "using natural convection for heat distribution"— letting the heat rise naturally from its collection point into your house (or storage). A thermosiphoning system looks superficially like an active solar air system (see page 24) because some of its components are the same: the collectors are panels mounted outside the house, and the heat is stored in a bin of rocks under the house.

But instead of moving the heat by mechanical means, a thermosiphoning system places the house *above* the collector to allow natural heat distribution by convection currents.

The collector is a frame supporting one or two layers of glass over a black-painted rectangular metal plate or wire mesh (see Drawings 45 and 46). It is mounted several feet below the house, facing south and sloped to catch the low winter sun (for slope angle see ''A Site for Sun-catching,'' page 38). Above the collector, the house's basement contains a huge bin of rocks. As the sun heats the collector's metal plate, it also heats the air between glazing and plate,

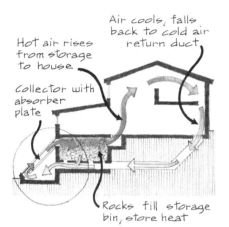

Drawing 45: Thermosiphon air heating system

Natural heat circulation occurs when hot air rises from collector into rock bin. If house needs heat, ducts open, hot air rises naturally from rocks up through house. Meanwhile, convection pulls cooler, heavier air down through vents on house's north side, back through ducts to collector.

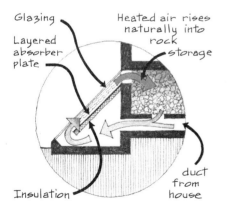

Drawing 46: Thermosiphon air collector

causing it to rise up through ducts into the storage bin (and, when needed, from storage through ducts into the house above), establishing a natural convective heat distribution.

Due to the different levels of collector, storage, and house, the natural convection currents are strong enough to bring heat all the way from the collectors to the house's upper levels. (See ''Getting into Hot Water,'' page 28, for thermosiphoning water heaters.)

When the sun is not out, the cold collector can be closed off from house and storage with dampers. Only the hot storage rocks and the house will then be involved in the convective loop, with cool room air returning to the storage, rather than the collector, to be heated once again.

In summer, the top of the collector can be vented to the outdoors to prevent the accumulation of unwanted heat in the storage and house.

Active: Panels, Pumps, and Fans

An active solar heating system is like a furnace with the sun as its heat source. Like a furnace, it consists of mechanical components, but the hardware is different: panels of metal and glass to trap the sun's heat, a water tank or a rock bin to store it, and pipes or ducts to convey it wherever it is needed.

It's called an active system because it uses thermostats, fans, pumps, and valves powered by a small input of electrical energy. These devices drive a heat-transfer fluid from panels to storage, and distribute heat throughout the house.

The two basic kinds of active systems are distinguished by the different heat-transfer fluids they use: air or liquid. These system types will be discussed in detail later in this chapter. First, we'll explore the general characteristics of an active solar heater.

Active solar heating begins with the solar heat **collectors.** These panels are usually mounted in rows on the house roof, though other locations may be chosen for better solar exposure. Like the south windows of a passively heated house, the panels must face within 20° of true south for maximum sun exposure—the closer to true south they face, the better their performance will be. They must also be tilted so they're nearly perpendicular to the low winter sun

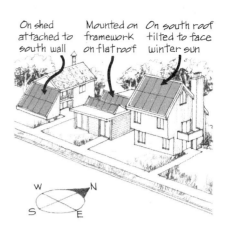

Drawing 47: Locating active collectors

(the tilt angle is calculated according to your latitude—see page 38).

The heat collectors must be free from shade in winter from neighboring trees, buildings, or hills; even a small area or period of shading during the day can cut down on their efficiency noticeably. Calculating the number and size of panels necessary to heat a house is complex, involving such variables as size and total heat loss of the house, local climatic conditions, availability of sunshine, and so on. For more details, see page 35. (Wherever the collectors are mounted, there should be some means of reaching them easily for cleaning and maintenance.)

The most common and least expensive kind of collector panel used for home space heating is called a **flat-plate collector.** Its functioning component is a metal plate, called an **absorber plate** because its function is to absorb sunlight and convert it to heat. The plate may be made of aluminum, copper, steel, or a combination of these metals. It is painted a flat (nonglossy) black or coated with a highly effective absorbent micron-thin **selective surface** to increase the amount of solar energy absorbed and decrease the amount of energy lost by reradiation. Selective surfaces collect heat so much more efficiently than paint that, in most climates, using them can obviate the need for a second layer of glass on a collector operating at normal house-heating temperatures.

(Continued on next page)

They are applied to the absorber plate by such technical methods as electroplating—to get a durable, high-quality selective surface you'll have to pay considerably more than for paint.

Housing the absorber plate is a shallow wooden or metal box covered with transparent glass or fiberglass (fiberglass or the more expensive tempered glass are less susceptible to breakage). Through the ''greenhouse effect'' (see page 9), this enclosure helps trap the sun's heat in the collector. The glazing may be single or double, depending on the climate the collector must withstand.

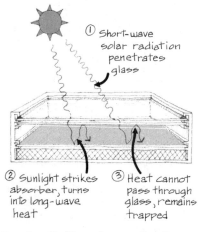

① Short-wave solar radiation penetrates glass

② Sunlight strikes absorber, turns into long-wave heat

③ Heat cannot pass through glass, remains trapped

Drawing 48: "Greenhouse effect"

If winters are very cold, as in the northern states, double or even triple glazing is necessary to minimize heat loss through the cover when the absorber plate builds up high temperatures. But if the climate is mild and heat loss is not a great problem, as in California or the Deep South, single glazing is preferable because it admits more sun.

The glass or plastic must be sealed to the housing box tightly enough to prevent rain leaks, wind or dust penetration, and heat loss, but flexibly enough to allow for the different expansion and contraction rates of the collector's different materials as they respond to daily temperature swings. Otherwise you risk glass breakage, broken seals, and misshapen collector components.

Between the absorber plate and the back of the housing box, a layer of heat-resistant insulation prevents heat loss through the box back. (For more sources covering your solar collector options, turn to the bibliography, page 93.)

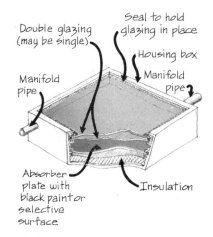

Double glazing (may be single)

Seal to hold glazing in place

Housing box

Manifold pipe

Manifold pipe

Absorber plate with black paint or selective surface

Insulation

Drawing 49: Components of a flat-plate collector

Even on a freezing winter day, bright or moderately filtered sunlight can heat the absorber plate of an active collector up to 200° F. (On summer days, when the fluid is not circulating in the collectors, ''stagnation temperatures'' can reach 400° F., so the collector must be designed to withstand such intense heat.) In the morning, as soon as the collectors are 10° to 20° warmer than the

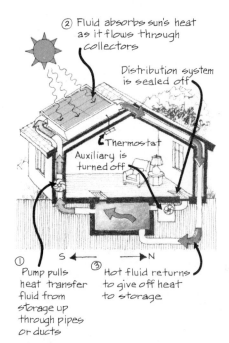

② Fluid absorbs sun's heat as it flows through collectors

Distribution system is sealed off

Thermostat

Auxiliary is turned off

S ← → N

① Pump pulls heat transfer fluid from storage up through pipes or ducts

③ Hot fluid returns to give off heat to storage

Drawing 50: Heat movement: collectors to storage

storage container in the house below, a **differential thermostat** senses the temperature difference and turns on a pump or blower to circulate the **heat-transfer fluid** through the collectors.

The fluid moves along the hot absorber plate, earning its name by picking up heat from the plate and conveying it through insulated pipes or ducts, back to the heat storage area. (The heat-transfer fluid usually enters the collectors at the bottom and leaves at the top, following heat's natural tendency to rise.) As long as the collectors are hotter than the storage, the fluid will continue to circulate from storage to collectors and back, getting hotter and hotter. The recirculating fluid can build storage temperatures up as high as 150° to 165° F. or more.

When the sun goes down (or disappears behind a cloud) and the collectors begin to cool, the thermostat detects the moment when the collectors' temperature drops close to that of the storage, and shuts off the pumps or blowers.

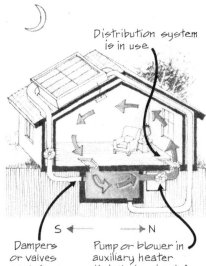

Distribution system is in use

S ← → N

Dampers or valves seal storage off from cold collectors

Pump or blower in auxiliary heater distributes heat from storage throughout house (auxiliary itself is turned off)

Drawing 51: Heat movement: storage to house

Meanwhile, the house has a thermostat to determine its own heat needs, just like a conventionally heated house. When heat is needed, the thermostat turns on another set of pumps or fans, which draw heat from the storage to warm the house proper. (Heat distribution from storage to house can differ between air and liquid systems—see pages 24 and 25.)

The storage capacity of an active system provides from 1 to 3 days' heat for the house without any input of heat from the collectors. This storage carryover of heat protects against long winter storms or cloudy periods, when the lack of direct sun prevents heat buildup in the collectors and replacement of heat in the storage.

During prolonged stormy periods, heat in storage is depleted; auxiliary heater comes on to heat house through same distribution system

Drawing 52: Backup heat

A prolonged cloudy period will deplete the heat in the storage. (Storage temperatures below about 90° F. are not usually high enough to be useful for space heating.) At such a time, an active system becomes unable to fulfill the heating needs of

a house, and an auxiliary heating system must supply the necessary heat until the sun returns for long enough to recharge the storage with heat. (Some of your auxiliary heating options are described on page 31.) Because of high initial costs and space requirements, it is rarely cost-effective to design a system with a long enough carryover to provide a house with 100 percent of its heating needs. Most full-scale systems are designed to provide 50 to 80 percent of a house's heating needs.

In order for an active system to perform well, the size of the storage must be matched to the total collector area. Other factors that affect the percentage of heat derived from an active solar heater include the house's ability to retain heat (through insulation, weatherstripping, double glazing, caulking, and window insulation) and the heat retention of the pipes (or ducts) and the storage, both of which must be heavily insulated. One of the most complex factors in the performance of an active solar heater is the efficiency of the collectors in gathering available sunlight.

Efficiency varies markedly from one model of collector to another. To calculate collector efficiency, you must consider many factors: the light transmission of the glazing, the light absorptance and emittance (see glossary, page 93) of the absorber plate's surface, heat conduction of the absorber plate's metal and efficiency of its exchange of that heat to the heat-transfer fluid, flow rate of the fluid along the absorber plate, operating temperatures of the collector (the higher the operating temperature, the lower the collector's efficiency), and quality of the collector's insulation. For sources covering such calculations, consult the bibliography on page 93.

The initial cost of an active solar system is unquestionably high compared to that of a conventional heater. In fact, on an initial cost basis alone, active solar heaters cannot be called economically competitive. Once it's running smoothly, however, an active space heating system can usually recoup the initial cost difference in 12 to 20 years, just through the savings on heating bills. (For the moment, solar heaters are more competitive with costly all-electric heating than with the currently less expensive gas or oil heat in most parts of the country.)

The main components of an active system—the panels and storage,

together with their installation— constitute the biggest expense of the system. Pumps, fans, thermostats, and controls add to the cost, as do heat exchangers (see "Liquid systems," page 25) and the continuing operating cost of running the machinery. But since most pumps and blowers used in an active heater are small—1/16 to 3/4 horsepower—they cost relatively little to run.

Because their collection and storage components don't require the aid of house windows, walls, or floors, active solar heaters need not dictate the architectural style of the houses they heat. (A whole active solar system could be erected in your backyard and operate entirely independent of the house, feeding heat to the house like an old-fashioned central steam-heating plant.)

For flexibility, an active solar heater's components can be arranged in a number of different configurations and proportioned in different sizes to suit the needs of various house types, ranging from the conventional to the highly unconventional. This adaptability not only makes such systems appealing to the prospective homebuilder, but it also makes active solar heaters one of the most practical solar retrofit options for existing homes.

Living in an active solar home is much like living in a conventionally heated home, once you have adjusted to the visual impact of the collector panels on the roof and become used to the tank or rock bin in the basement. Since the heat is stored in one location, sealed off from the rest of the house, its distribution can be thermostatically controlled so heat is used only when it is needed, just as with a conventional heater.

Such control over heat flow also means that heat can be directed to specific rooms anywhere in the house, giving flexibility in room placement and permitting the use of closed zoning in the interiors of active solar homes (see "Zoned heating," page 9).

Day-to-day operation of an active system, like that of a conventional heater, is relatively carefree: there are no vents, doors, or windows that control heat flow.

(Continued on next page)

In most active systems the auxiliary heater is connected to the thermostats that control the solar heater; the auxiliary turns on automatically whenever the solar heat storage is used up, maintaining a uniform comfort level all winter long, no matter what the weather or time of day. (Nevertheless, these systems do require a certain amount of monitoring of delicate controls and checking for mechanical malfunction.)

Air systems. An active solar heater that uses air to pick up solar heat from the collectors and deposit it in the storage is known as an air system. Air systems have their own distinctive forms of collector, distribution system, and storage, even though their basic operation mirrors the general description already given.

To pick up the collectors' heat, the air must be blown in front of and/or behind the absorber plates in an even stream by a fan, coming in contact with as much of the plates' surface as possible. The absorber plates themselves can take a number of forms (see Drawing 53); their common characteristic is the textured surface, which serves to agitate the flow of passing air, making it swirl against every inch of the hot metal and increasing its opportunity to pick up all the heat available in the collector. All types of absorbers are painted flat black or coated with a selective surface to increase their absorptance of the sun's rays.

When the heated air leaves the top of the collectors, it is carried to storage in large, well-insulated ducts. These ducts constitute one drawback of air systems: they take up substantial space in a new home and are difficult to retrofit into an existing home not designed for active solar heating.

On a positive note, however, air systems operate at relatively low temperatures—and the lower the temperature the less heat lost from the collector, and the more efficiently the air can pick up the collector's heat. In addition, air systems run no risk of leaking liquids through your roof or walls or into your cellar. Air systems are relatively easy to maintain, and they present no problems of corrosion or winter freezing, and few summer overheating difficulties in the collectors.

On days when the house is chilly inside but the sun is shining, an air system can direct heated air straight from collectors through ducts to the house, without detouring through the storage at all.

Once interior heating needs have been met, the hot air is directed down to the bin full of rocks that constitutes an air system's heat storage. The auxiliary heater is usually either electric resistance heating in the ducts or a gas or oil furnace.

The rock bin is the most common type of storage. The bin, usually in the basement, is a boxlike concrete container, poured along with the foundation when the house is built and then insulated heavily on the outside with rigid-board insulation or spray-on urethane foam. (Bins may also be made of insulated plywood or sheet metal.) Into the container go dense rocks, such as smoothly rounded river rocks, granite, or basalt. Their size, 1 to 4 inches in diameter, is chosen to balance uninterrupted air flow with adequate heat exchange—hallmarks of a good rock storage.

Rock storage has the disadvantage of being extremely bulky; 100 to 400 pounds of rock are needed per square foot of collector—about 2½ times the storage volume needed for equivalent heat storage in a liquid system.

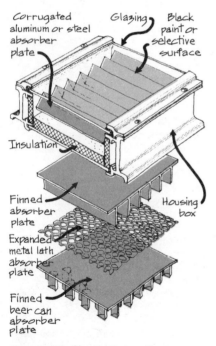

Corrugated aluminum or steel absorber plate
Glazing
Black paint or selective surface
Insulation
Finned absorber plate
Housing box
Expanded metal lath absorber plate
Finned beer can absorber plate

Drawing 53: Types of air collector

If rock storage is well stocked with heat, hot air from collectors can be blown directly into house. Then bypassing storage, house air is cycled back to collectors

S ← → N

Storage closed off from collectors, house

Drawing 54: Air system: collector-to-house cycle

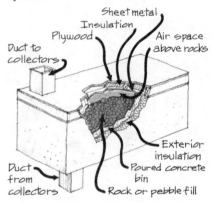

Sheet metal
Insulation
Plywood
Duct to collectors
Air space above rocks
Exterior insulation
Poured concrete bin
Duct from collectors
Rock or pebble fill

Drawing 55: Rock heat storage

Eutectic salt storage, the other alternative for an air system, takes up much less room than rocks (a particular advantage for retrofits). These salts have the property of melting at a relatively low temperature (usually around 80° F.), such as that of air heated in solar collectors. As they melt they absorb large quantities of the air's heat. Then, when cool air

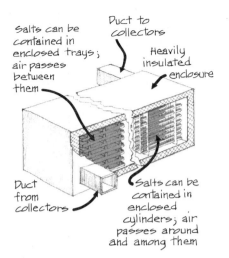

Salts can be contained in enclosed trays; air passes between them

Duct to collectors

Heavily insulated enclosure

Duct from collectors

Salts can be contained in enclosed cylinders; air passes around and among them

Drawing 56: Eutectic salt heat storage

from the house blows by the salts, they solidify again, releasing their stored heat to warm the air. Two and one quarter pounds of Glauber's salt (the most commonly used eutectic salt) can hold as much heat as 132 pounds of water, heated 1.8° F., or 690 pounds of rock, heated 1.8° F.

Liquid systems. If an active system uses a liquid to carry heat from collectors to storage, then the system is known as a liquid system. Like air systems, liquid systems have their own style of collectors, storage, and distribution.

The absorber plate for the flat-plate collector of a liquid system is a metal surface containing closed channels through which the liquid can flow, picking up the metal's heat as it goes. Liquid absorber plates are of several different types. The most common type is the **tube type**: a single metal sheet with metal tubes either clamped tightly into the plate itself or bonded to its top or bottom surface with heat-conductive solder or adhesive.

Another type of absorber plate is the **tube-in-plate**: two sheets of metal, bonded together to form integral passages within the plate through which the liquid passes. The tubes in both types are carefully spaced and sized to balance efficiency of heat collection with cost of materials. In addition, the tubes are

arranged in mathematically determined flow patterns on (or in) the plates, so as to maximize heat transfer from plate to liquid, assure a uniform flow of liquid through the tubes without any sudden drops in pressure, and prevent "hot spots" caused by the inability of the passing fluid to remove heat from areas of the plate too remote from the tubes.

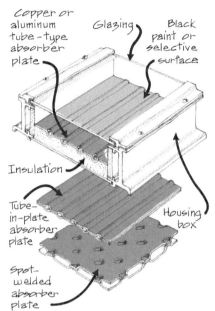

Copper or aluminum tube-type absorber plate

Glazing

Black paint or selective surface

Insulation

Tube-in-plate absorber plate

Housing box

Spot-welded absorber plate

Drawing 57: Types of liquid collector

When mounted on the roof for heat collection, most liquid collector panels are laid out in horizontal rows, connected at bottom and top with horizontal manifold pipes. These pipes carry liquid to and from the collector array and from row to row of the panels.

Flat-plate liquid absorber plates are made from copper, aluminum, and stainless or galvanized steel. Each metal has its merits and drawbacks. Steel is inexpensive but only a moderately good conductor and prone to corrosion by the liquid passing continually through its tubes. Aluminum, also inexpensive and a better conductor, is very prone to corrosion. Corrosion and rust inhibitors can be added to the heat-transfer liquid, but they only reduce, not solve, the problem.

Copper, on the other hand, is far more resistant to corrosion than the other metals, and it is an excellent conductor—but it is very expensive.

A frequent compromise is to bond or clamp copper tubes to steel or aluminum plates, so as to get the

greatest collection efficiency for the least cost. The only hitch to this solution is that such bonding is difficult, due to the different expansion rates of the metals as they heat, which can break the bond and cause poor conduction of heat from plate to liquid. Corrosion is also a serious danger wherever two different metals come together.

Another variety of liquid collector differs from the types described above and avoids some of their problems. This is the **trickle-type** collector, invented by Harry Thomason. It has the simplest of absorber plates, made of a sheet of corrugated aluminum or galvanized steel, painted flat black. The heat-transfer liquid (generally plain water) flows openly down the channels in the metal, picking up heat without being enclosed in

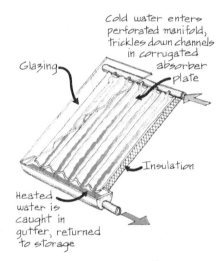

Cold water enters perforated manifold, trickles down channels in corrugated absorber plate

Glazing

Insulation

Heated water is caught in gutter, returned to storage

Drawing 58: Trickle-type collector

a tube at all. Only the collector glazing shields the water from the outside air.

Some experts say that this open flow is inefficient and that it permits erosion of the painted surface of the absorber plate and causes heat losses through condensation on the underside of the glazing. But the trickle-type collectors are inexpensive and relatively simple to maintain, and they perform well at the moderate temperatures needed for home space heating. They are also free of the freezing problems that threaten other types of liquid collectors.

(Continued on next page)

At the opposite technological extreme from the trickle-type collector is the **concentrating collector.** This collector bears little resemblance to a flat-plate collector. Instead, it consists typically of a single, black-painted metal pipe (the absorber) with a **parabolic reflector** of polished aluminum mounted lengthwise beneath it like a mirror-lined trough. The reflector focuses the sun's rays onto the pipe, concentrating so much energy on it that the temperatures produced in the heat-transfer fluid flowing through the pipe are much higher than those produced by flat-plate absorber plates.

But in order for the parabolic reflector to bounce light onto the pipe properly, it must pivot to follow the sun across the sky during the day, in order to keep the sun shining directly into the center of the parabola. Also, the collector uses only clear-day direct radiation, whereas flat-plate collectors can also use diffuse radiation on hazy days to produce heat.

The machinery needed to produce this slow daily rotation of the reflectors is expensive to buy and run (since it requires electricity), and it offers more opportunity for breakdowns. Besides these disadvantages, most experts generally agree that the high temperatures produced by a concentrating collector are not necessary (or, considering the high initial cost, economical) for home heating. (Solar cooling is another matter—see page 33, "Sun-powered Cooling.")

Linked with the choice of collector type is the choice of liquid to be used as the heat-transfer fluid. Here—unless you live in a rare area where temperatures never plunge below 32° F.—the main problem to consider is freezing.

If you use water, and the temperature outside drops below freezing while the water is standing in the collectors, the water will freeze—bursting the absorber channels just as it does the plumbing of an old house. Therefore, to use untreated water you must have a **drain-down** system, in which a thermostat at the collectors senses an approaching freeze. The thermostat signals the circulating pump to drain all the water out of the collectors and back down into the storage to await a thaw.

The trickle-type collector works well in avoiding freezing problems because it drains down naturally and completely as soon as the water's circulation stops.

A liquid system that uses water and drains down to avoid freezing is called an **open-loop** system, because the water runs directly from the collectors into the storage tank, and back to the collectors (see Drawing 60).

One frequent solution to the freezing problem is to use antifreeze—ethylene glycol, propylene glycol, or mineral oil. Antifreeze can be used in solar collectors just as in car radiators, allowing the liquid to remain in the collectors without risk of freezing. But antifreeze has two disadvantages. First, it breaks down and becomes corrosive unless replaced every year or two. And second, building codes require that if your solar space heating system also heats your household water, you can use antifreeze in the system only if it is a **closed-loop** one. A closed-loop system uses a **heat exchanger**—a long, loosely coiled copper tube or a finned device like a car radiator, double-walled for extra protection—immersed in the storage tank to keep the toxic antifreeze separate from the

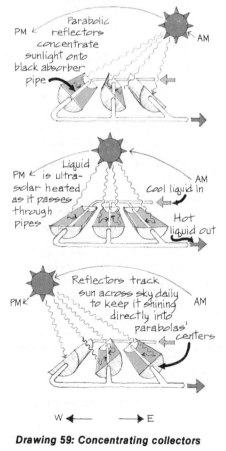

Drawing 59: Concentrating collectors

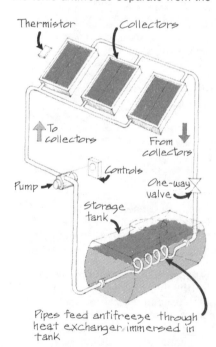

Pipes feed antifreeze through heat exchanger immersed in tank

Drawing 61: Closed-loop system

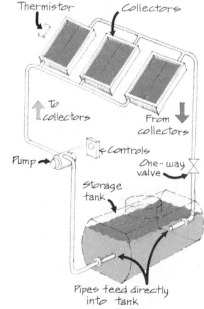

Pipes feed directly into tank

Drawing 60: Open-loop system

water (see Drawing 61). The heat exchanger makes the closed-loop system less efficient than the open-loop type, because less of the heat from the collectors is transferred to the storage water; this condition decreases both the efficiency of collector operation and the efficiency with which the storage is warmed.

The collectors are the most complicated aspect of an active liquid heater. Once out of the collectors, the hot liquid follows a simple path (the shorter the better, to avoid heat loss) through pipes to the storage tank, usually in the basement. The piping, valves, and small pumps that control and distribute the liquid involve fairly standard plumbing techniques. In contrast to the ducts for an air system, they consume little space and can easily be retrofitted.

However, careful plumbing and a close watch are necessary to avoid leakage on the roof around the collectors and in the plumbing inside the house.

The storage of a liquid system is relatively compact; only 1 to 4 gallons of water are needed for every square foot of collector. The water is stored in a tank made from any of several materials. Glass-lined steel tanks are one ponderous option, rather difficult to get into a house.

Another option is a concrete tank (possibly poured along with the foundation, much as a rock storage bin is). Then, because concrete will eventually leak, the inside of the tank is lined with thick polyethylene, which resists high temperatures and provides protection against leaks. (The liner must be replaced after a period of years.) This concrete/liner combination is hardy and relatively low in cost, but difficult to retrofit.

Better adapted for retrofitting (and fine for new houses) are the new lightweight, noncorrosible, but quite expensive fiberglass tanks, some of which have been designed specifically for solar heating systems. All storage tanks must be very heavily insulated.

The final step in an active liquid heating system is delivering heat from the storage tank to the house. With an air system this is easy, because the hot air from storage can be fed directly into the rooms. Liquid systems are more complicated.

One effective way is through a radiant floor heating system (see Drawing 63). Some conventionally heated houses have hydronic radiant floor heating systems that retrofit very neatly with liquid solar systems.

More common are radiant baseboard heaters (see Drawing 64), which offer the same level of comfort as electric baseboard heaters. (In fact, electricity or a gas-fired boiler serves as a good auxiliary heat source for both radiant floor and baseboard heaters.)

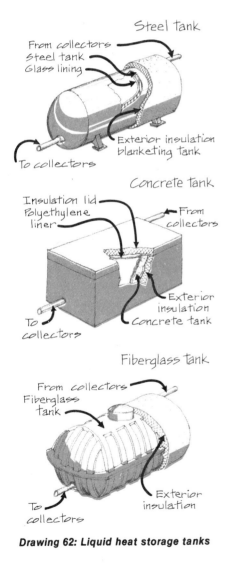

Drawing 62: Liquid heat storage tanks

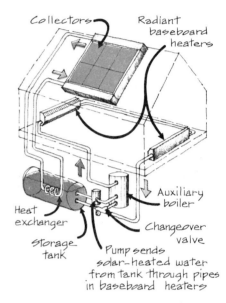

Drawing 64: Radiant baseboard heat distribution

The third and most prevalent option is forced-air heating. In order to use a forced-air heater with a liquid solar system, you must install another heat exchanger—water to air—in the return air duct from the house to the auxiliary gas or oil furnace. While solar-heated liquid passes through the heat exchanger, the furnace blower pushes house air past it to pick up the heat (see Drawing 65 on page 28). Should the storage be too cool to heat the house, the furnace will come on and heat the house air instead.

(Continued on next page)

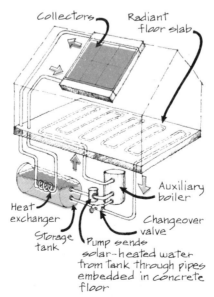

Drawing 63: Radiant floor heat distribution

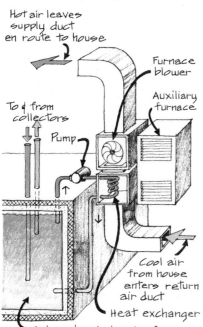

Hot air leaves supply duct en route to house

Furnace blower

Auxiliary furnace

To & from collectors

Pump

Cool air from house enters return air duct

Heat exchanger

Solar-heated water from storage tank is pumped though heat exchanger in supply duct; blower extracts heat, blows hot air to house

Drawing 65: Forced-air furnace heat distribution

Active/Passive: The Hybrids

Many people adapt and combine elements of both active and passive solar heating systems to meet their needs. Some have found that passive and active solar heating can complement one another very effectively. These combination active/passive solar systems are called "hybrid" systems. For example, a sun-tempered home that draws in the sun's direct heat through well-placed south windows and holds it in with berms may be ideally suited to active solar heating, because its heating needs are minimal, and because it already faces the winter sun. An active system can provide a longer carry-over if combined with passive elements so that the active storage holds its heat for night use, while the house is heated passively during the day.

A south-facing greenhouse can provide a welcome boost for an actively heated house, and also give you a flourishing year-round garden. Many owners of passive solar homes find active domestic water heaters (discussed below) a real boon.

Hybrid systems enable homeowners to benefit from both types of heating, assembling the house that suits their tastes and needs from the passive and active devices they choose. Those who have created their own hybrid combinations of solar systems have generally found the resulting houses varied, intriguing, and highly successful in terms of comfort.

Getting into Hot Water

We've discussed the first major heating need of the average home—space heating. Now let's turn to the second—domestic water heating. The typical family of four uses up to 80 gallons of hot water a day. In a conventional home, the hot water is produced in an electric or gas-fired water heater, accounting for up to one-third of the house's total heating bill.

Here's where solar heating can reap remarkable savings: up to 90 percent of your hot water needs can be fulfilled by solar heating. And a well-designed solar water heater can pay for itself in 4 to 10 years—any hot water you use after that is virtually free. Solar water heaters are far more compact and less expensive than space heaters: they usually have two to five collector panels, often with the conventional water heater acting as both storage and backup heater. Easily retrofitted onto almost any home, a solar water heater works the year around. Solar water heating is more efficient than solar space heating because it operates in summer—when the most sun is available—as well as in winter.

Three kinds of solar assistance are available for your domestic water heating: passive preheating, active full heating, and passive full heating.

Passive preheating. Preheating means taking advantage of the sunlight already available in your house to warm the domestic water on its way from the water main to the conventional water heater. Preheating is easiest if your house already has some form of passive heating system, such as a greenhouse, Trombe wall, or direct-gain windows (see pages 16–21), where sunlight is

already collected, and where a preheater might be placed for heat gain.

Here is one of the more inventive (and successful) ideas for a passive preheater. The tank of an old water heater is taken out of its insulating case, painted black, and placed directly inside south-facing windows or suspended by sturdy bolts and chains in the south-facing skylight of a bathroom, laundry room, or kitchen.

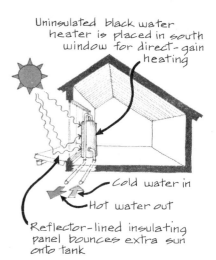

Uninsulated black water heater is placed in south window for direct-gain heating

Cold water in

Hot water out

Reflector-lined insulating panel bounces extra sun onto tank

Drawing 66: Passive water preheater

There the sun streams in on it all day long. An insulating panel, lined with polished aluminum, may be attached to the window's exterior to seal heat in at night, swinging down during the day to reflect extra sunlight in the window (see Drawing 66). Or, if the tank is under a skylight, a curved, polished reflector may be mounted on the skylight's north side to reflect extra winter sun down onto the tank below. By nightfall, the water inside the tank is hot. (If no insulation for window or skylight is provided, most of the heat will be lost by morning.)

Full passive heating. For full-scale passive water heating, there are thermosiphoning water heaters (for an explanation of thermosiphoning, see page 20). In all outward respects, a thermosiphoning water heater resembles a small, active liquid space-heating system (see page 25); the difference is that it has no pumps. Instead, the water circulates by natural convection, rising and falling in response to the sun's heat just as air does.

To initiate this convective current, the collectors are mounted with their tops *below* the bottom of the well-insulated tank into which they feed (see Drawing 67).

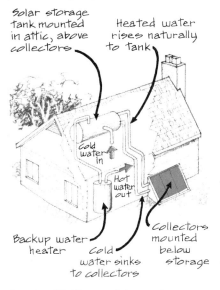

Drawing 67: Thermosiphon water heater

An extra storage tank between collectors and water heater allows water to recirculate through the collectors numerous times for higher temperatures before going to the water-heater tank, where it may be drawn off for household use.

This system is an open-loop one, using plain water. To prevent night or wintertime freezing, the collectors must be drained when the temperature drops (see page 26). (If the system is a full-scale active one, it can be allowed to feed back a little hot water from the storage to the collectors.)

If you use antifreeze as your heat-transfer liquid to avoid freezing problems, or if your corrosion-prone collectors necessitate addition of a corrosion inhibitor to the water, building codes (and common sense) require a double closed-loop system to prevent poisoning your drinking water. You can do this in either of two ways: with two heat exchangers in your storage tank—one for the hot antifreeze from the collectors, the other for tap water—or with a double-walled heat exchanger.

Full active heating. If you decide on a full-scale active solar water heater, so you can mount the collectors on the roof and place the tank at the more typical floor level, you'll need pumps, valves, and automatic controls to circulate the water or antifreeze mechanically (see Drawing 68).

Though most such systems are custom-built, a market is developing for "off-the-shelf" solar water heaters—preassembled package systems that include collectors, tanks, thermostats, pumps, and piping. With some plumbing and wiring skills, you may be able to install one of these yourself on a house with an appropriate south-facing, tilted or flat roof. Even a shed or garage roof will do: only 1 to 2 square feet of collector area is needed per gallon of water to be heated. If the roof is flat, you'll have to mount the collectors on a rack to tilt them (see page 38).

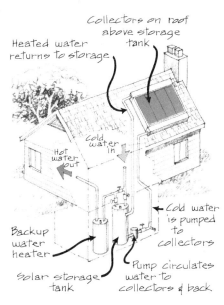

Drawing 68: Active solar water heater

These packages are less expensive than custom-built models, but they are not simple to install, requiring quite a bit of expertise and derring-do (and a good dose of patience) on your part for a safe plumbing installation. (Be sure before buying that the instructions are adequate, or that the seller will help you if they aren't.) If you are not an experienced plumber, it's best to have your system installed by an expert, even if you have to pay more.

If you're already planning an active air or liquid space-heating system, you should consider including a domestic water-preheating or heating system (see Drawings 69 and 70).

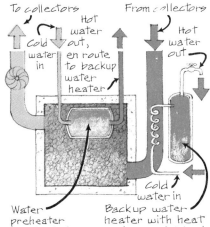

Drawing 69: Domestic water preheat in air space-heating system.

Two options exist for preheating domestic water with an air-type space-heating system: a preheater tank buried in rock storage, or a heat exchanger installed in return duct from collectors. Water is cycled through for heating, on its way to backup water heater.

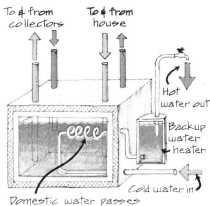

Drawing 70: Domestic water preheat in liquid space-heating system

Pools: A Place in the Sun

Solar heating and swimming pools are a most compatible combination for several reasons. First, the cost of conventional heat for a swimming pool is so high (simply because the pool loses so much heat from its broad, exposed surface every night and cloudy day) that it makes solar costs look more inviting. Second, the swimming season corresponds to the season of high, bright sun and long days, so plenty of solar heat is available when most needed for the pool. And finally, since pools do not need a tremendous amount of high-intensity heat, and because they act as their own heat-transfer liquid and heat storage, heating a pool requires fewer and less complicated solar components than heating a house. (Spas and hot tubs are another matter—see below.)

Heat conservation for a swimming pool is the first important step toward solar heating. It can be accomplished with a pool cover. These come in a variety of styles; some are moved mechanically, others manually. Any cover that fits snugly and completely over the water will help to hold the pool's heat at night and on cloudy days (see Drawing 71).

Solar pool heaters are active liquid systems (see page 25). The liquid collector panels may be mounted on a

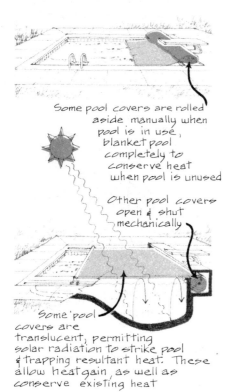

Some pool covers are rolled aside manually when pool is in use, blanket pool completely to conserve heat when pool is unused

Other pool covers open & shut mechanically

Some pool covers are translucent, permitting solar radiation to strike pool & trapping resultant heat. These allow heatgain, as well as conserve existing heat

Drawing 71: Pool covers

south-facing house, garage, or pool-house roof, or on a nearby earth bank or wood framework constructed to hold them. To heat the pool, a pump cycles pool water through the collectors and back to the pool again (see Drawing 72). A differential thermostat turns on the pump when the collectors are hot enough to benefit the pool. Most systems also include a regular pool filter and an auxiliary heater.

Collectors
Thermistor Control

Pump — Pool filter — Control valve determines whether water goes to collectors or to auxiliary heater

Cool water to collectors

Hot water into pool

Drawing 72: Solar pool heating

Since the pool water is used in the collectors, its chemical balance (pH) must be carefully maintained to prevent corrosion in the collectors (if the absorber plates are metal).

The various kinds of solar pool collectors differ in price, longevity, and output temperature. The simplest, least expensive, and shortest-lived are black plastic panels. They are often made of polyvinyl chloride (PVC) tubing, either in sheets of tube-in-plate design (see page 25) or coiled in circles—see Drawing 73. They operate efficiently in putting out fairly low temperatures (all that is necessary for pool heating) and last 5 to 10 years before the plastic begins to break down. They have no housing box or glazing and are mounted directly on the roof. Their temperature never gets high enough to damage the roofing.

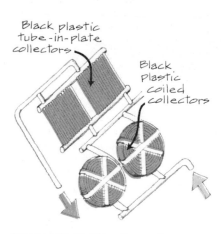

Black plastic tube-in-plate collectors

Black plastic coiled collectors

Drawing 73: Plastic solar pool collectors

Full-fledged liquid collector panels—metal absorber, housing, glazing, and all—are also used for pool heating, though they're expensive compared to the other type. The reasons for using such panels include their durability and efficiency as well as their usefulness for additional solar space heating. In some cases, the same collector panels can be used for summer pool heating and winter house heating. (Some systems use the pool itself as a winter heat source and summer heat sink for their solar-assisted heat pump—see page 32.)

A solar pool heater can extend your swimming season by 2 to 4 months, depending on your geographical location. In California a season can run from April to October, whereas in New England it's more likely to last from June to September.

Solar Spas and Hot Tubs

Unlike swimming pools, spas and hot tubs require quite high temperatures—110 to 115°F. For this reason a solar hot-tub system resembles a domestic water heating system (see page 28) more than it does a pool heating system.

For a typical tub 6 feet in diameter and 4 feet deep, supplied with the necessary insulating blanket or lid, about 130 to 160 square feet of single or double-glazed liquid collector panels will provide sufficient heat for year-round hot-tub soaking. (The unglazed PVC panels used for pool heating will not produce high enough temperatures for hot-tub heating.)

The insulating blanket turns the hot tub into its own heat storage container; no extra tank is necessary. Unless the system is a thermosiphoning one (see page 20), it will require a pump to circulate water from hot tub to panels and back, as well as thermostats and valves. (In freezing climates a drain-down system must be provided to prevent freeze damage to the collectors—see page 26.)

The flow rate controls the temperature of the water coming from the collectors. A faster flow rate than would be used with a domestic hot-water system keeps the water from scalding the bathers as it feeds directly back into the tub.

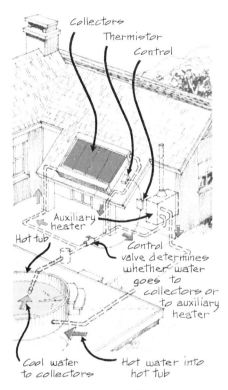

Drawing 74: Solar hot-tub heating

For people who want both a spa or hot tub and a pool, a single solar heating system can cover the needs of both. This system must use glazed metal collectors (as opposed to PVC) to obtain high temperatures, and heating priority must be given to the

spa or tub, with heat going to the pool only when the spa is already hot enough. About 200 to 260 square feet of collectors are necessary to provide enough overflow heat to warm the pool adequately.

If you don't have a pool but are planning to add a solar hot tub, consider adding solar domestic water heating as well (or vice versa). The two systems are quite similar in design, so they're easy and economical to combine.

In this case again heating priority should be given to one system, with overflow heat "preheating" the other. For example, if the hot tub is your main concern, you would circulate the hot-tub water through the collectors until the tub reached the proper temperature. At that point the differential thermostat would stop the hot-tub pump and direct the water-heater pump to begin circulating your domestic water to the collectors for heating. (Alternatively, a heat exchanger from the water heater could be immersed in the hot tub, and domestic water cycled through to pick up excess heat from the hot-tub water.) This combination system requires six to eight panels; a drain-down system must be provided in freezing climates (see page 26).

To be cost-effective, a solar hot tub must be used every day, just as domestic hot water is. If you are only a weekend tub user, it is cheaper, at current gas prices, to use conventional heating, switching it on for the weekend and leaving it off during the week.

The Backup: Auxiliary Heating

One of the great economic drawbacks of solar space and water heating is the fact that during long, cold, cloudy periods, the solar heater just cannot provide all the necessary heat for the house. With no other form of heat to fall back on, you could find yourself in serious trouble when your solar heater runs out of stored heat in the midst of a snowstorm. Even in areas with consistently mild winters, some additional heat will be needed.

In areas where firewood is readily available, some small houses can get by comfortably using high-efficiency woodstoves as backup, and equipping their fireplaces with heat-recovery devices for an extra boost (see Drawings 75 and 76).

Drawing 75: Woodstove backup

The woodstoves are the descendants of the Franklin stove; they burn wood slowly and thoroughly at a very high temperature. They are freestanding to allow heat to radiate into the room from all surfaces of the stove.

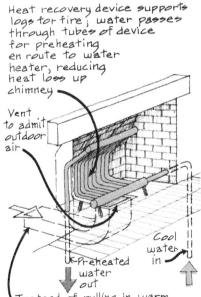

Drawing 76: Fireplace heat-recovery device

(Continued on next page)

In addition to woodstoves, some houses in mild areas, such as California and Florida, use a scattering of electric baseboard heaters. These are relatively inexpensive to buy, though their dependence on electricity makes them expensive to run. But in most areas of the U.S., laws require a full-size conventional gas, oil, or electric heater as auxiliary to a solar heater, despite the added expense of seldom-used machinery.

Some auxiliary heaters run completely separately from the solar heater, coming on only when the thermostat indicates that the solar heat storage is depleted. All passively heated houses have this type of auxiliary heater arrangement. Other auxiliary heaters keep the heat storage of an active solar heater at an even temperature when the sun alone cannot maintain the storage heat (though some designers argue that this is inefficient).

The heat pump. One alternative to the conventional types of heat backup for an actively heated house is the electric heat pump, a device that seems to produce more energy (heat) than it consumes (electricity). In actuality, the device just gathers up sun-generated heat energy already present in even very cold winter air and moves it into the house. Though a small input of electricity is required to power this process, the heat pump requires only 1 Btu of electricity to produce up to 3 Btus of heat (a "coefficient of performance" of up to 3 to 1, under optimum conditions), whereas a regular electric resistance heater produces only 1 Btu of heat for every Btu of electricity it uses.

The heat pump looks and acts just like an air conditioner that can turn itself around to heat. The system is easier to understand if you remember three principles of physics:
—Heat energy flows naturally from a warm place to a cooler place.
—When caused to expand, a gas becomes colder.
—When compressed into a tighter space, a gas becomes hotter.

Like an air conditioner, the heat pump consists of a closed loop of refrigerant gas that circulates through two finned heat-exchanger coils, one inside the house and one outdoors. Each heat exchanger has a blower fan to move air across its surface.

The refrigerant is volatile—it boils from a liquid to vapor at low temperatures. An electric compressor is on one side of the refrigerant loop between the two coils; a refrigerant expansion valve is on the other (see Drawing 77).

For winter heating, the liquid refrigerant passes through the expansion valve, rapidly expanding (and thus cooling) as it approaches the outdoor coil. The refrigerant is colder than outdoor winter air (the "heat source"), so atmospheric heat moves into it through the coil, boiling the liquid into a vapor. The vapor is then squeezed down by the compressor, becoming denser and hotter—it is hotter than the room air when it passes through the indoor coil. The compressor's mechanical energy has been converted to heat energy, boosting the heat pulled in from outdoors.

At the indoor coil, the blower carries heat away from the hot gas and into the house (the "heat sink") through a normal ducting system. As it loses heat, the gas condenses to a liquid and moves through the expansion valve to repeat the process.

For summer cooling, a thermostat essentially reverses the flow of re-

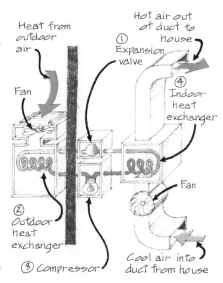

Drawing 77: Heat pump heating cycle
Liquid refrigerant is expanded to cold liquid and pushed outdoors (1). Heat from outdoor air moves into cold liquid in heat exchanger, boiling liquid to warm vapor (2). As vapor is compressed, it becomes hotter and is pushed indoors (3). Heat transfers to air blown past heat exchanger in duct; heat loss causes vapor to condense to liquid (4).

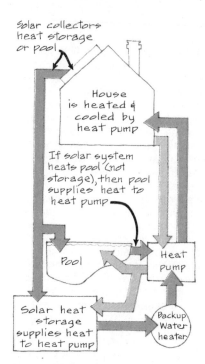

Drawing 78: Solar-assisted heat pump
Heating cycle, phase 1 (using storage only, not pool): heat is delivered from storage directly to house; heat pump and backup are turned off. Phase 2 (if storage is too cool for direct heating, or pool is used as heat pump's heat source): heat pump extracts heat from storage or pool and delivers it to house; backup is turned off. Phase 3 (if storage or pool is too cool to be used as heat pump's heat source): heat pump extracts heat from backup water heater and delivers it to house. Cooling cycle: heat pump extracts heat from house and dumps it in storage or pool.

frigerant so that it captures heat indoors and carries it outdoors to disperse it, leaving the indoor air cooler.

A heat pump's auxiliary heating capacity can be combined with an active solar heater to the advantage of both. Though the heat pump can work completely on its own, its performance drops sharply when its heat source is colder than 35°F. (freeze-up problems may also occur with the outdoor coil). But solar heat storage is not very useful below 80° for forced air heating, or 90° for radiant heating. However, if the heat pump is designed to use the solar heat storage as its heat source (see Drawing 78), it can use sun-warmed air or liquid down to 35° or 40°F. and still give close to peak efficiency and heat output. (For greatest efficiency, the solar system should heat the house directly whenever its storage is hot enough, bypassing the heat pump.)

This arrangement is particularly good for cold but clear winter areas. (In mild climates, a solar-heated swimming pool can serve as winter heat source and summer heat sink.)

Because of its built-in air-conditioning capacity, a heat pump is considerably more expensive than an ordinary auxiliary heater. It is most useful in climates where both its cooling function *and* its economical heating function are required for year-round comfort. Though it is not generally competitive with current natural gas prices, it can compete economically in many areas with oil, propane, and electric heating and air conditioning. In addition, its high coefficient of performance makes the heat pump a valuable aid in scaling down the demand for electricity during high demand periods. It may end up being an important savings for solar homeowners who would otherwise pay high penalties for their reliance on electric radiant heaters as bad-weather auxiliaries.

Sun-powered Cooling

Paradoxical as it may seem, the sun can be a great aid in cooling your house. This is easier to understand if you think of the sun as an *energy* source rather than just a *heat* source. With air conditioning costs rising as fast as electrical heating costs, the sun's cooling-energy potential is an important resource. Solar cooling ranges from age-old, simple methods to complicated techniques that are still on the frontiers of technology.

Natural ventilation. One of the simplest and most widespread forms of solar cooling is **natural convection ventilation.** Many houses designed for solar heating use variations on this kind of cooling: windows are opened near the floor on the north side of the room and near the ceiling on the room's south side. During the day, as the air in the room heats up, the warm air rises and escapes out the open south windows, drawing cooler air after it through the north ones (see Drawing 79).

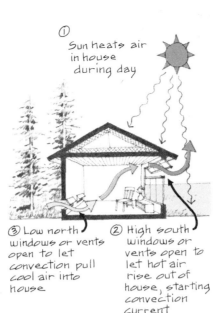

① Sun heats air in house during day
③ Low north windows or vents open to let convection pull cool air into house
② High south windows or vents open to let hot air rise out of house, starting convection current

Drawing 79: Convection cooling

As we have seen in "Passive: The House as Heater" (page 14), many passive solar heating systems have a cooling mode as well, involving either natural ventilation, using the sun's heat to set up a cooling convection current, or nighttime radiation.

Nighttime radiation. Active air systems can use nighttime radiation for cooling purposes as well, without additional elaborate equipment. On a cloudless night with relatively low humidity, warm objects will radiate their heat to the cooler night sky. In areas where such nights are common, the function of a standard ac-

tive air system can be reversed, so that the heat storage loses its heat through nighttime radiation. This is done by venting the rock storage directly to the cool night air (see Drawing 80). Cooling can also be obtained by using a blower to pull cool night air through the storage rocks, storing "coolth" for daytime circulation in the house.

Absorption chillers. Homeowners considering an active liquid system have another cooling option: a device known as an **absorption chiller.** Solar-powered absorption cooling moves a technological giant step beyond nighttime radiation. Rather than simply releasing accumulated heat, it actively extracts heat from the house air. The process is complicated, but like the heat pump (or your kitchen refrigerator) it involves a closed loop of refrigerant. (If you have a heat pump, however, that should provide all necessary cooling at lower cost—see page 32.)

In order for an absorption cooling unit to work with any degree of efficiency, it must have an input of solar-heated liquid no cooler than 180°F. This rules out most flat-plate collectors, since they cannot consistently produce such high temperatures. Some—endowed with highly efficient selective surfaces on their absorber plates and at least two layers of glazing—can do it, but the high temperature level of operation means low efficiency in collector performance and thus requires a large collector area. By far the preferred form of liquid collector for an absorption cooler is the concentrating collector (see page 26), because of the high temperatures it produces at peak efficiency.

Solar-powered absorption cooling for residential use is just emerging from the experimental stages. Because of its complex equipment and high-performance collector panels, its initial cost is still extremely high. (If the cooling collectors can be used for winter heating as well, the cost seems a little less exorbitant.) But if air conditioning is very important for comfort in your climate, absorption cooling may well do the job.

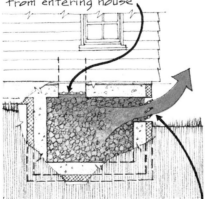

Dampers close duct from rock storage to house to keep heat from entering house

Dampers open rock storage to outdoors to let heat escape to night sky. (Or fan may pull cool air in through rocks to cool them.)

Drawing 80: Night cooling of rock storage

Can the Sun Work for You?

Guidelines for planning your own solar heating system

Planning your solar heating system is no simple matter. A lot of calculations and thoughtful decision-making are involved in determining the solar potential of your home or site, designing your system, selecting its components or materials, and installing them properly.

The first big step toward solar heating is deciding what kind of system is most suitable for you. But it's still a long hike from there to the installation itself, with many calculations and decisions to be made along the way.

Begin by reconnoitering your site. Analyze the demands its latitude, climate, and layout will make on your solar design. If you want to retrofit an existing house, you must also determine the limitations that existing conditions impose on the house's solar potential.

Moving along, you come to design choices in profusion: Do you want an active or a passive system? High performance? Low cost? Will you hire a contractor or build it yourself? And do you want solar-assisted cooling as well? Building code and deed restrictions must be taken into account, and other legal problems may arise as well.

Beyond these considerations lies the cost. Is solar heating really economical? Can you find financing? What about tax breaks?

To assist you in searching for sound answers and solid judgments as you embark on your solar installation, we offer the guidelines in this chapter. There is no single sure path to a satisfactory solar installation: the technology is still too new, federal and state regulations and system performance standards for the solar industry have not yet crystallized, and the fast-growing solar market is still wide open to self-styled "solar designers" of sometimes questionable training and expertise.

However, using the advice included in this chapter, and adding some thoughtful deliberation and homework, you should end up with a solar heating system well suited to your life style and your heating needs.

Climate and Site

The first chapter discussed the importance of a rapport between a whole house and its climate and microclimate (see page 7). As your viewpoint narrows to the solar heating system itself and you begin to plan a solar installation, you must return once again to a consideration of your climate and site—this time as they affect the size, type, and performance of a solar heater.

Climate Considerations

The two most important climatic factors affecting a solar heating system are the demand for winter heat and the corresponding amount of sunshine available. (If cooling is a concern, then the temperature range and prevalent humidity level during the summer cooling season are also important factors.)

Who needs the most heat? Any house, no matter how tightly constructed and well-insulated, will lose heat on a cold winter day. The colder your climate is and the longer the cold weather lasts each year, the more heat your house will lose in the course of a heating season—and, consequently, the more heat your heating system must provide to make up the difference and keep the house comfortable.

To calculate the climate's demand on your heater, you must first determine the number of degree-days in your heating season. A **degree-day** is a measurement used to compute the difference between the outdoor temperature on a given day and a fixed standard of 65°F. (the point at which most people's furnaces turn on). Every degree of difference between 65° and the average outdoor temperature on a given day equals 1 degree-day. This means that if the outside average temperature is 35°F. for one day, you have accumulated 30 degree-days on that day. If the temperature averages 40° the next day, the degree-days for that day will number only 25. The total degree-days for the 2 days will be 55.

Monthly degree-day records for many localities are available from the National Weather Bureau. Degree-days for a complete heating season in the U.S. can range anywhere from 600 (in Tampa, Florida) to 10,800 (in Anchorage, Alaska). The higher the total figure, the more heat your heater is obliged to produce during the heating season.

The second measurement to pinpoint in estimating the climate's demands on your heating system is the average temperature low point anticipated each winter in your climate region. This temperature is called the **design temperature,** because your heating system is designed to maintain comfort in your house even when the temperature

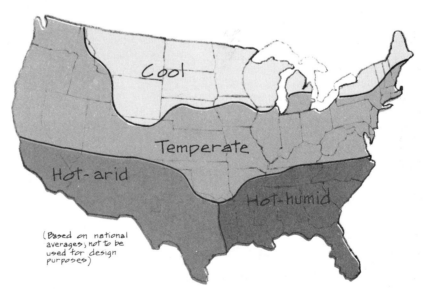

Drawing 81: Climate regions of U.S.

outside is within its lowest range.

Using the design temperature and degree-day measurements for your climate, and taking into account the external surface area of your house and the heat transfer capacity of all the materials with which it is built and insulated, you can make a series of calculations to estimate the total heat loss from your house every heating season.

If cooling is also needed to keep your house comfortable, you must calculate the climate's demands on your cooling system much as you did for your heater. To calculate the number of degree-days in your cooling season, measure how much hotter each summer day is than 80°F. (the point at which most people turn on their air conditioners). The design temperature for a cooling system is the average temperature high point expected in a year.

These heat loss (or gain) calculations are standard ones that can be done for you by a solar architect or designer or a heating and cooling engineer. If you wish to make your own, consult the bibliography on page 93 for further sources of information. For detailed climatic data, write the National Weather Records Center in Asheville, North Carolina, for a copy of the U.S. Department of Commerce's *Climatic Atlas of the United States.*

Who gets the most sun? The crucial aspect of climate for a solar heating system is the availability of sunshine. The system's heat output depends directly upon the amount of solar radiation striking the collector area from month to month.

Sunshine is measured in two ways: clear-day insolation and the percentage of possible sunshine. The clear-day **insolation** (incident solar radiation) is the amount of sun that strikes a surface at a given latitude and tilt angle, through direct radiation from the sun, diffuse radiation from the sky, and reflected radiation from the surface's surroundings. (Drawing 82 shows the clear-day insolation of a horizontal surface, reflected radiation not included.) The **percentage of possible sunshine** is the percentage of time during the average year that the sun is bright enough to cast a shadow across a surface, divided by the number of hours the sun is above

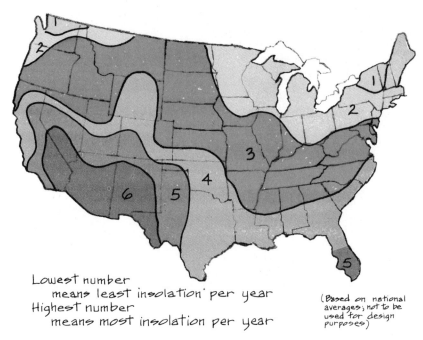

Lowest number
    means least insolation per year
Highest number
    means most insolation per year

(Based on national averages; not to be used for design purposes)

Drawing 82: Annual mean daily solar radiation

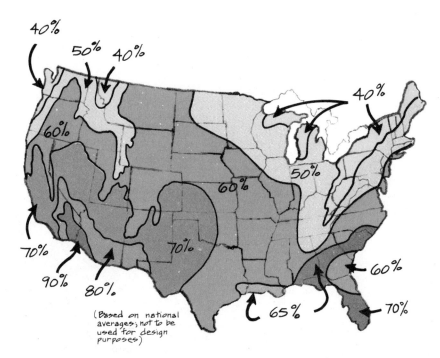

(Based on national averages; not to be used for design purposes)

Drawing 83: Annual mean percentage of possible sunshine

the horizon. Your local weather bureau (or the National Weather Records Center in Asheville, North Carolina) can provide you with weather maps and monthly (or even daily and hourly) data on both insolation and percentage of possible sunshine. Their records are accurate enough to allow calculations for most low-temperature solar heating design.

By combining the clear-day insolation data for your latitude and the data on the percentage of possible sunshine, you will learn not only how much sunlight will strike your solar heater's collector area on the optimum clear day, but also the amount you'll gather on hazy or cloudy days, and the percentage of the daylight hours during the heating season that's likely to be clear or cloudy. These figures will help you to determine the amount of heat you can expect to obtain from your solar collector area at crucial times of the year.

When you compare the amount of sunshine available to you at these important times with the data on your house's heat demand at the same times, you'll get a much clearer picture of the performance you're likely to get from your solar heater depending, of course, on the efficiency of your solar heating system. A complete appraisal of the availability of sunshine in your climate can be done by the same solar consultant who does your heat loss calculations (see page 35). Or you can do it yourself; for sources of information, consult the bibliography on page 93.

The optimum weather conditions for solar heating are bright sun on the coldest days of the year, so the collector can gather plenty of heat just when it's most needed. The areas of the U.S. where these conditions prevail most often are Colorado and the Southwest, which are therefore the optimum areas for solar heating. Unfortunately, in many other areas of the U.S. (especially the north and northeast), cold weather and high heat demand are often accompanied by cloudy skies, making solar heating somewhat less reliable than in the southwest. However, if you look at the insolation and percentage of possible sunshine data for these areas, you may be surprised how much useful sunshine is available for heating through the course of the winter. Though the percentage of your total heating demand your

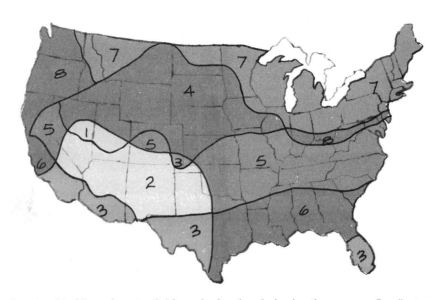

Drawing 84: Climate's potential for solar heating during heating season: *Excellent sunshine is combined with high heat demand in area 1, moderate heat demand in area 2, and low heat demand in area 3. Good sunshine is combined with high heat demand in area 4, moderate heat demand in area 5, and low heat demand in area 6. Fair sunshine is combined with high heat demand in area 7 and moderate heat demand in area 8. Map taken from TRW, Phase 0 Study, "Solar Heating and Cooling of Buildings," National Science Foundation, 1974.*

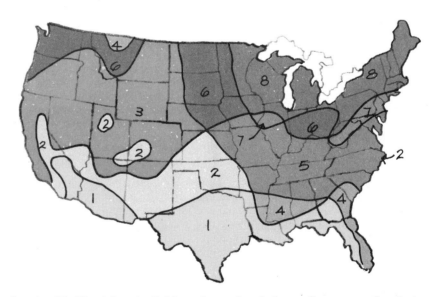

Drawing 85: Climate's potential for solar cooling during cooling season: *Excellent sunshine is combined with high cooling demand in area 1, moderate cooling demand in area 2, and low cooling demand in area 3. Good sunshine is combined with high cooling demand in area 4, moderate cooling demand in area 5, and low cooling demand in area 6. Fair sunshine is combined with moderate cooling demand in area 7 and low cooling demand in area 8. Map taken from TRW, Phase 0 Study, "Solar Heating and Cooling of Buildings," National Science Foundation, 1974.*

solar heater can provide may be lower in Minnesota or Connecticut than in New Mexico, it can still make a sizable contribution, especially if the house conserves heat adequately and the solar system performs efficiently.

A Site for Sun-catching

Whether you are designing a new house to be equipped with solar heating or are considering a solar heating addition to your existing home, two siting factors are important: the southern exposure and (especially if you plan an active system) the latitude of your site.

We discussed southern exposure at some length in Chapter 1 (page 7), but some of the points bear repeating because they are crucial to the success of a solar heating system. Orientation is first: Any solar collector area—passive windows or active panels—should face within 20° of true south (which differs from magnetic south—the "south" indicated by a compass). In climates where mornings are regularly hazy or misty, a slight southwest orientation may even be an advantage, simply because more sun is available in the afternoon than in the morning. If early-morning warm-up is desired, or late-afternoon overheating is a problem, a slight southeast orientation may be preferable.

Orienting your collector area to the south is only part of the battle for good sun exposure. You must also be very careful that no shading of the collector area occurs, particularly in winter during the optimum sun-collection hours, approximately 9:00 A.M. to 3:00 P.M. Some summer shading is permissible if you're not trying to heat a pool or your domestic hot water or power a solar absorption cooling device (see page 33) with the shaded collector area.

Any collector, passive or active, combines heat gain from the sun with heat loss to the cold air. If part of the collector is shaded, that part will

When collectors are installed, site may appear shade-free

Several years later, neighbors' trees and/or house may have grown, rendering collectors inoperable due to shading

Drawing 86: Shading hazards

gather very little heat from the sun, but it will continue to lose heat to the air and will even rob heat by conduction from neighboring sun-struck collector areas.

Remember that shadows which are short in summer may be disastrously long in winter. Any object—a tree, a house, or whatever—to the south of your collector area is a potential liability. Trees can grow and neighbors' houses can acquire second stories, causing unanticipated shading problems. Try to locate your collector area as far away from present or potential shading problems as possible. (Shade cast by telephone poles or trees with thin, bare branches is not serious.)

The latitude of your house has an important effect on an active solar heating system because it determines the noon sun's angle above your horizon (see page 7). The noon sun's angle, in turn, determines the optimum tilt angle of your collector area. Your collector should be as

close as possible to perpendicular to the sun at whatever time of year you need the sun's heat most. If winter heat is your goal, tilt your collector to meet the low winter sun. A good rule of thumb is to add 15° to your latitude to obtain the optimum tilt angle from the horizontal for winter heat collection. If you need heat to power a summer cooling device, subtract 15° from your latitude to get the op-

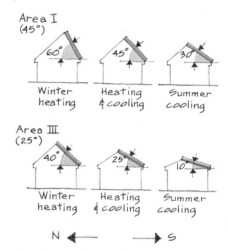

Drawing 87: Roof angles for optimum solar exposure

timum collector tilt angle. If you need both space heating and cooling, domestic water, and/or pool heating, try a tilt angle equal to your latitude. A collector's tilt can vary within 15° of optimum without much loss of efficiency. (Some collector supports allow you to adjust the angle seasonally.)

Many passive systems lack the flexibility to tilt to the optimum angle for heat collection: walls and windows are for the most part vertical, and roof ponds horizontal. However, their sun intake can be enormously increased with the addition of reflectors. For a vertical collector, snow lying on the ground beneath the collector area is one good reflector. If you have no snow, white gravel can take its place. Insulating shutters, raised above (in the case of roof ponds) or lowered to

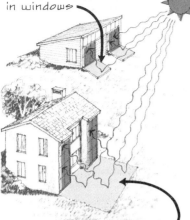

Drop-down insulating panels are reflector-lined to bounce more sun in windows

Light gravel area (or snow) reflects extra sun onto vertical collectors for improved heat gain

Drawing 88: Reflectors for vertical collectors

the ground below the collector area and lined on the inside with polished metal plates or foil, can better perform the same function: to reflect extra sunlight onto the collector area. At a latitude such as New England's (40° to 45°), a vertical collector area combined with reflectors can be at least as effective as a collector area at the "optimum" tilt.

For more details on evaluating the solar potential of your site and climate, see the bibliography on page 93.

The New and the Old

A person who's building both a house and a solar heating system at the same time has different solar design concerns than someone planning to add a solar heater onto an existing home. The former has only climate and site specifications to constrain his imagination, whereas the latter must also find ways to adapt the demanding solar components to his existing house—unless he wants to do a major remodeling job. These two solar planning situations each deserve a closer look.

Starting from Scratch

If you are lucky enough to be designing and building your home to include solar heating, many choices are open to you. Your choice of system type is relatively unrestricted: passive, active, or a hybrid of the two. You can situate and design your house for optimum orientation toward the winter sun and (if your system is to be active) for the appropriate roof angle. (If it suits the contour of your site better, permits a magnificent view, or is necessary to comply with your building code setback regulations, your house may be turned as much as 20° southeast or southwest—see page 38.)

Your landscaping can be tailored to permit the winter sun to penetrate to your collector area. Regardless of the system type you choose as the actual solar heater, the house can be designed thoughtfully for maximum heat conservation and for use of the sun's heat directly through well-placed windows for sun-tempering. The interior of the house can be zoned appropriately to use the solar system's heat efficiently. Space for the massive heat storage area can be planned into the house's design, along with space for controls and pipes or ducts. For an active solar system, easy access to the collector area can be provided, and the routing of the heat-transfer fluid from collectors to storage to house can be planned for simplicity and proximity to prevent undue heat loss in transit.

The advantages of a house originally designed to include solar heating cannot be underestimated. Added together, they have an enormous effect on the solar system's performance, by conserving the precious resources of heat in every possible way. They allow the components of an active or passive solar heater to be used to maximum advantage, often permitting the system to be smaller and therefore less costly. The majority of solar homes in the U.S. today were designed especially to contain solar heating systems.

The Retrofit

Designing a solar space-heating system to fit an existing house is much more difficult than designing a house together with its solar system, because the options are much fewer.

Solar pool heaters, which operate at relatively low temperatures and thus do not require the efficiency of a space heating system, and domestic water heaters, which require few panels and a small storage area, are relatively flexible; they adapt to differing conditions better than a solar space heater.

Not all existing homes are adaptable for solar space heating. They may have high heat losses (rendering the heat demand too large for an economical solar heating system to fulfill). They may have the wrong orientation, suffer deep winter shading, or simply lack the space for the solar components. If you're interested in the possibility of retrofitting a solar heater onto your present home, the first thing to do is analyze your home's solar potential.

A home must meet several criteria to be eligible for a solar retrofit. It must have an area suitable for heat collection: large enough to gather adequate heat for space heating, with a southern orientation and unobstructed exposure to the winter sun.

In the case of a passive heating system, this collection area must be a part of the house, and be aligned with the storage mass, so that the mass can absorb the sun's direct heat. The significant costs and dif-

Solar greenhouse heater can be retrofitted onto south, west, or east unshaded wall. Some south exposure is necessary; existing windows, doors in house wall are an advantage

Drawing 89: Solar greenhouse retrofit

ficulty of the structural changes necessary to add most passive systems make such retrofits unlikely choices for most existing houses, unless you add a solar greenhouse onto a south-facing wall, convert an existing masonry wall to be a Trombe wall, or plan extensive structural remodeling.

As mentioned in Chapter 1 (page 23), active systems have a flexibility that makes them better adapted to retrofitting than many passive systems. The collectors, for example, may be installed on a roof with appropriate slope and orientation (and even the slope can be modified by using racks to support the collectors). But if your roof doesn't have these attributes, or if there are obstructions (dormer windows, plumbing vent stacks, chimneys, or the shadow of a neighboring house,

for example) which break up the potential collector area inconveniently, you may find another location offering the necessary 60 to 600 or more square feet (depending on the type of solar heating—domestic water, pool, or space—you intend to add) of unobstructed south exposure. A vertical area such as a wall or fence is one potential spot; another is the roof of a garage, pool house, or shed. Freestanding collector areas can be constructed on your property, either leaning against a south wall or completely separate from the house. And freestanding collectors can be mounted on a flat roof or a roof with a very low pitch (though wind and, in some locations, earthquakes can pose problems, requiring very sturdy structural support for such collectors). If freestanding collectors are mounted on a roof, any chimneys within 10 feet of the panels must extend 2 feet above the tops of the collectors, and plumbing vents must be 6 inches taller, to prevent safety hazards. A collector area can even be divided into several separate smaller areas, though this may cause piping or ducting complications and should be avoided, if possible.

All of these options involve reconciling yourself to a highly visible, even space-consuming, set of solar collectors. Some obstructions—TV antennas, drainage downspouts, some plumbing vents, or even windows or skylights—may have to be moved to make a convenient collector area. (Be sure to consult a builder or plumber before you make any structural or plumbing changes. Check also with your city or county building or planning departments—see "Building Codes and Legal Rights," page 43.)

The next criterion for active solar retrofitting is space in your home for the distribution system and the sizable volume of storage.

The heat distribution for an active liquid system is relatively easy to retrofit (provided the plumbing is done carefully to prevent leakage), because the pipes and pumps involved take up little room. Air ducts and fans for an active air system demand much more space; they also require a very direct route from collectors to storage if the system is to perform efficiently. This ductwork

may be difficult to retrofit unless your house already has extensive ducting for a conventional forced-air furnace, or you are willing to have large heavily insulated ducts running openly through your rooms.

With regard to storage: either a liquid or an air system's storage area demands quite a bit of space—though once again, the air system's rock storage will require three times the volume of space a liquid system's tank is likely to need.

The ideal spot for either form of storage is the basement, if you have one that's large enough. However, it is difficult to get a tank or rock bed of the proper size *into* most base-

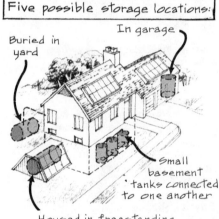

Five possible storage locations:

Buried in yard

In garage

Small basement tanks connected to one another

Housed in freestanding shed on which collectors are mounted

Drawing 91: Active retrofit: storage

ments without major construction (though several small water tanks, connected together, can be used for liquid heat storage, enabling you to move them through a cellar door one by one, and there are some water tanks designed for retrofitting, which can be disassembled so as to fit through a door, then reassembled and lined inside with vinyl water-proofing).

Other possibilities for locating the heat storage include burying it in the yard (a rather expensive proposition

Five possible collector locations:

Mounted on freestanding shed in garden, facing south

Mounted on east or west-facing roof; turned to face south

Mounted on shed against south-facing wall (or, with reflectors, on wall itself)

Mounted on south-facing garage roof

Mounted atop freestanding trellis in garden

Drawing 90: Active retrofit: collectors

involving complicated piping or ducting), installing it in a garage (if the garage is large enough to accommodate it), or even building an outdoor shed around it. The closer the collectors are to the storage, the better, to avoid unnecessary heat loss. For freestanding collectors, one idea is to use the framework that supports the collectors to house the heat storage container as well, keeping the two close together. (If this is done, the storage must be doubly well-insulated.)

Finally, having decided on an appropriate area for the collector and a space for heat distribution and storage, you must link your conventional heating equipment with the solar components to provide both solar heat delivery to the house and a backup source of heat for cloudy weather.

For example, if you wish to link solar components to a domestic water heater, you must know where the cold water main enters the conventional water heater, so that you can intercept it and send the cold water up to the collectors first. For pool heating, you must determine where to intercept the circulation of water from the pool in order to pump it from the pool through the filter to the collectors, and back through the conventional heater to the pool.

As for space heating, the kind of conventional heating you have determines how you can combine it with the type of active solar heater you plan to add (see "Air systems," page 24, or "Liquid systems," page 25) so that it provides both solar heat distribution and auxiliary heat.

If you're interested in combining a solar heating system with a heat pump (see page 32) in your retrofit, you'll need space in your furnace or boiler room to put the heat pump equipment.

A detailed analysis of your home's solar potential can be made by a qualified solar architect, engineer, or designer, or you can do it yourself. See the bibliography on page 93 for further sources of information.

Once you have ascertained that a solar retrofit is feasible for your house, you have one more job before embarking on your solar installation.

You must make your house as heat-tight as possible—by adding insulation, weatherstripping, double-glazed windows (and perhaps movable window insulation), and insulation for your water heater and for your heating ducts or pipes. Many homeowners are interested in solar retrofitting mainly for economy—saving money in the long run. If that is your concern, take note: the energy conservation steps just named are the least expensive way to reduce the size, and therefore the initial cost of a solar heater by reducing the heat demand on it. It is decidedly cheaper, in general, to *conserve* energy in the first place than to *replace* it using a solar heating system.

Guidelines: Getting Started on Solar

If the solar outlook for your home or building site seems promising after you've appraised its solar potential, and if you've decided on the system type (or hybrid of types) best suited to your needs, you're ready to start the design process. The first question that comes up at this stage is: Who's going to design and build your solar heater?

Hire an Expert or Do It Yourself?

Designing a solar heating system—active or passive—is no easy business. In Chapter 1 you've already had a glimpse of the technical complexities involved in the design of a passive house's structure or an active system's components. You've also met the complexities of heat loss calculations, availability of sunshine, optimal siting for solar heating, and the added difficulties of retrofit in this chapter. Solar heater design encompasses all these details, calculations, measurements, and compromises, plus many specifics we're unable to cover here. To do your own designing, you must prepare yourself by reading widely in the solar literature available (at critical points in this book, we refer you to the bibliography on page 93 for reliable, in-depth sources of information). Discuss your thoughts

and questions with willing solar designers and homeowners. Enroll in a course on do-it-yourself solar design and construction offered by your local community college or adult education program, or by a local solar organization, such as your regional Solar Energy Association (to find the nearest Association, consult our bibliography, under "Solar Energy Associations and Information Centers," page 93). Be sure to compare the information you acquire from different books and authorities; opinions vary tremendously even on technical questions, and no single answer is the only correct one. Look for the solutions that best fit your particular needs.

As you embark on the design itself, you may be able to get assistance with the calculations. Computer programs that can perform heat loss calculations, size your system, predict its efficiency of performance, and run cost analyses are available from the federal "Sol-cost" program, some solar design firms, regional Solar Energy Associations, universities, and state energy commissions. You can use one for a fee. (Check the Yellow Pages under "Solar" or call the nearest university or state energy commission office.)

Once you've formulated your design plans, you'll need to hire a professional draftsman to make up the working drawings. Even if you do your own drawings, it is a good idea to have a solar architect, engineer, or designer evaluate your plans, calculations, and working drawings before you begin the actual installation. (You will have to pay a fee for such an evaluation, but the money is well spent as insurance against expensive and even hazardous errors in the details of your design.)

If you feel that designing your own system would demand too much time or too sophisticated a technical background in construction and engineering, you may wish to hire a solar architect, engineer, or designer to do all your calculations, design the system, and prepare the working drawings for you. The designer will consult you, just as a regular architect would, about your needs and preferences before and during the design work. Once the design is complete, the designer will provide you with a detailed cost analysis of the proposed system and any recommended heat conservation precautions.

The more you know about the kind of solar heating system you want when you hire the designer, the better control you'll have over the resulting plans. In addition to system type, you should be able to tell the designer your priorities: What can you afford? Is low cost important to you? Would you prefer a highly efficient, high-performance system to one that is simpler in technology but turns out fewer Btus of heat per square foot of collector? Do you want a fully automated system, or one you will operate manually? Do you want a large space-heating system, to provide 80 to 90 percent of your heat, or would you prefer a less ambitious system, providing 50 to 60 percent of your heat? Should your house look conventional or unconventional? The designer will work with your priorities in mind, though some compromises may be necessary between your wishes and the demands of your climate and site. The cost for the designer's services may constitute up to 10 percent of the total cost of your solar installation.

Be careful in choosing your architect or designer; every solar expert has an individual style and technical expertise. Look at examples of the designer's completed work, if possible (or at least photographs of it) to see if it matches your taste. Check to see if the designer is familiar with the solar system types you prefer. Has he or she worked with them before? Then ask for references of homeowners whose systems he or she has designed; find out if the homeowners are satisfied with the work. Also check with the Better Business Bureau and with any local solar offices, such as your regional Solar Energy Association.

To avoid any chance of fraud, it's a good idea to find out how long a solar designer has been in business in your area, and what his or her financial references are. Solar heating is such a new and wide-open field, and it requires such a sizable financial investment, that it's wise to be very cautious in selecting a solar "expert." There are already plenty of cold solar collectors and hot lawsuits due either to solar firms that ran into financial difficulties or to fly-by-night "solar heating specialists" who vanished when their systems proved inoperable.

Who Will Do the Installation?

Once your system is designed, you come to the question of who's going to do the installation. If you've hired a solar architect, engineer, or designer, you'll probably want to let him or her find a reliable contractor, choose the components (if the system will be an active one), supervise the construction and inspect the completed work. If you've designed your own system, however, and are buying the components (perhaps even a solar heating kit) yourself, you are entering another round of decision-making that requires caution.

Choosing components. Because the solar industry is new, standards for equipment durability and performance have not yet been settled on an industrywide basis (though ASHRAE has developed some standards for flat-plate collector testing which have been adopted by the American National Standards Institute). As a result, the quality of a component may be hard to judge before you buy, especially since many manufacturers are new to the industry and have not yet established a reliable "track record."

The sophisticated technical nature of active solar components makes them more susceptible to mechanical imperfections than the more familiar, everyday materials used in passive solar systems. The list of active solar component and system failures compiled by disgruntled solar homeowners is a long one.

Sometimes the collectors are faulty: they spring leaks and drench the roof or allow hot air to escape; the glazing cracks as it expands, or the glazing seals break; the tubes in liquid absorber plates lose their bond to the plate.

High summer stagnation temperatures wreak havoc with poorly constructed collectors: The sun may heat collectors up to 400°F. in the summer when pumps or fans are not circulating the fluid to carry away the heat. At those temperatures, plastic covers will bend or melt; black paint can literally vaporize and rise to coat the glazing, or flake or melt off the absorber plate; in liquid collectors, solder can melt, causing bad bonds or leaky tubes; and, if not drained down, water or antifreeze can boil in the tubes, causing an explosion unless the system has a pressure release valve.

Even if the collectors prove durable under all conditions from hot to freezing, the other components— delicate thermistors (temperature sensors) and differential thermostats, valves, pumps or fans, plumbing, and ducts—can develop problems. Poorly made tanks can leak water; improperly designed and filled rock beds may not pick up and store heat adequately; and eutectic salts can cease to change phase (melt and solidify) to store and release heat unless chemically treated to prevent stratification of the salts' ingredients. Specialized controls for highly technical active systems are very susceptible to breakdown. Uninsulated ducts or pipes, vents that do not close tightly, valves that leak—all will rob heat from a system and cause its performance to be disappointing.

Since there is no reliable "seal of approval" to tell you which components are good investments and which are not, the responsibility for finding reliable equipment rests upon you. It is advisable to take some precautions before you buy. First of all, inform yourself on the technical requirements (and trouble spots) of the solar equipment you need by reading up on it beforehand. Then, comparison-shop: examine a number of different options and discuss their construction and function in detail with the seller.

When you think you've found the equipment you want, ask the seller for proof of the product's performance—preferably test results from a university, government, or independent private laboratory. Also ask about the warranty: its length (most cover 1 to 5 years, still far short of the 10 to 20-year lifetime predicted for most solar components and systems); whether it covers the full system or only a few components (if so, which components?); whether it covers servicing, parts, and labor in the event of a breakdown; and whether a broken component will be repaired on site or must be returned to the manufacturer. Find out specifically who will do the servicing. Don't settle for vague claims that "any electrician or plumber can do it"; it should be done by someone in your area who is reliable and readily available.

Check the seller's reliability. Does he have professional experience with the solar equipment he's selling? Has he been in the area long? What references can he supply—both from homeowners using his equipment and from financial institutions? Does he have an escrow account which will cover the warranty should his business fail before the warranty expires? Be alert for exaggerated performance claims; if the seller makes any claims about his equipment's performance, be sure to get them in writing before you purchase the equipment.

When you buy the components for a full heating system from one manufacturer, solar store, or solar designer, that source may be able to supply its own trained installers to do the construction work for you. This is a good arrangement, provided you're not interested in doing it yourself, because it gives you some assurance that the installers are experienced with solar heaters and are familiar with your system. If the installation is included in the terms of your warranty, you have some assurance that the manufacturer or designer will bear the responsibility for seeing that the system works properly once installed. This is especially important because, due to the newness of the technology, even a well-designed system using high-quality components is likely to develop minor problems during its early stages of operation. Most solar homeowners report that up to a year is required to "work out the bugs" and bring the system up to its predicted rate of performance.

Finding the right contractor. If, as is often the case, the manufacturer or designer has no trained installers, you will have to find a suitable contractor to install your solar system. You can do this by looking in the Yellow Pages under "Solar" or by recommendation from solar homeowners or local solar organizations.

Be sure that you deal only with a licensed contractor; obtain a written contract from him before the work begins. Get the contractor's references, as you did those of the architect or designer you worked with, and check out the quality of his past work. Try to find a contractor who is trained or experienced in solar installations. Many problems can arise from using a contractor who is inexperienced in solar heater construction or installation. Check the contractor's credentials with local solar organizations and with the Better Business Bureau. You are safest with someone whose reputation and financial background are solidly established in the area.

Being your own boss. Doing your own solar installation can save you a great deal of money. However, you shouldn't even consider doing it yourself, either using your own design or a do-it-yourself kit, unless you are an avid and experienced handyman. (If the system is to be an active one, you should be experienced with plumbing or sheet metal work and electrical wiring as well as basic construction.) Think twice before you let your enthusiasm commit you to doing all your own construction work: one or two errors could ruin your system and even cause structural damage to your house, costing you a great sum to repair it. Less drastic difficulties are also expensive: if you reach a technical impasse in mid-installation, you might need to hire a contractor to straighten out and finish off the job. Even if you succeed in completing the installation, it must function well for an extended period of time to be economical. A few seemingly minor mistakes in design or installation could bring increased trouble and cost in maintaining the system and reduce the system's level of performance. A do-it-yourself installation will save you money in the long run *only* if you are really equipped with the tools and expertise to do it well.

If you're anxious to cut down on costs, but honestly don't feel ready to tackle the whole installation yourself, consider doing just the preparatory work for the contractor, and perhaps assembling beforehand the components or building materials he'll need.

Building Codes and Legal Rights

As with any kind of major remodeling or building project, a solar heating installation requires a building permit and must conform to the building code and zoning regulations of your city or county.

Most jurisdictions have adopted one of the major national codes—the *Uniform Building Code* in the western states, the *Basic Building Code* in the middle Atlantic and northeastern states, the *Southern Standard Building Code* in the southeastern and southern states, or the *National*

Building Code in any region. These basic codes are usually amended to cover local building conditions. As yet, however, few codes have special sections applying to solar heating systems. Local building officials tend to adapt rules written for other situations to the solar system plans they review. If the department is lenient and favors solar heating, this lack of specific rules can work to the advantage of the homeowner proposing the system. However, if department personnel are unfamiliar with solar systems or concerned about them, they may cause difficulties for the homeowner by interpreting the existing codes too strictly.

You should be able to find a copy of the code at your local building department or library for consultation. In addition, several proposals for solar installation codes are presently being considered by state and local building departments, should you wish to research the rules that may apply to solar systems in the future. The best-known of these is the 1977 edition of the "HUD Intermediate Minimum Property Standards Supplement—Solar Heating and Domestic Hot Water Systems," developed by the National Bureau of Standards in connection with HUD.

To find out about the necessary building permit, call the local building or planning department. They will probably be able to provide you with a fact sheet showing such basic requirements as setback regulations, height restrictions, parking space requirements, and permissible percentage of lot coverage and percentage of window area.

If you're working with a designer, architect, engineer, or contractor, that person should apply for the building permits, supplying the necessary plans. If you're on your own, however, you may need to hire architectural or drafting help to produce the detailed scale drawings showing property line setbacks, topographic features, easements, floor layouts, electricity, plumbing, etc., necessary to apply for a building permit. You can shorten the draftsman's work time (and lessen the cost) by consulting the building, planning, and health departments yourself to determine all the factors that affect your plans. Don't hesitate to ask questions regarding your plans before you submit them for approval. (Depending on the type of system or scale of construction you're planning, you may need to apply for plumbing and electrical permits as well as the building permit.)

Application for each permit requires payment of a nonrefundable fee (usually a percentage of installation costs). Approval may take a day or several weeks.

In many cities and counties, architectural review boards as well as the building and planning departments review building plans. Their primary concern is usually the appearance of your system. Depending on their attitude, you may have to modify your plans to conform to their esthetic judgments.

You may need a variance. If you find, in examining the local zoning or planning ordinances, that your solar installation plans do not conform to them exactly, but you feel that the deviations from the restrictions are sensible and justified, you may apply for a variance. A common example of such a situation would be the addition of an attached solar greenhouse which oversteps the setback restrictions by a foot. The deviation may make sense, particularly if the lot is a narrow one to begin with, and if there is no other possible location for the greenhouse. In such a case, a variance should be sought.

Applying for a variance usually involves a written application (with fee) and a hearing where the ruling is made on whether or not to allow the variance. Exceptions to the zoning ordinances are often allowed, particularly in the case of solar heating, where a literal interpretation of the restrictions seems impractical or illogical, and where the variance will not be harmful to neighbors, adjacent property, or public welfare. Remember: Variances are as much a part of the law as restrictions; you have a right to present your reasons for asking that the rules be waived.

Sun rights. A few knotty problems in solar legislation remain to be solved. The worst of these is the question of sun rights. What happens if you install an expensive active system in your house one year, and the next year an apartment house goes up on the neighboring lot to the south of yours, effectively screening the low winter sun from your collectors? What recourse do you have if your next-door neighbor plants a tall yew hedge that shades your solar greenhouse in winter? The answer in most areas at present is, "Very little."

However, considerable study and debate among lawyers and national, state, and local legislators is being devoted to this problem. And in the meantime, there are a few things you can do.

First, check the zoning in your area for height restrictions before you build your solar heater. If your home is in a zone where only one or two-story single-family dwellings are allowed, and the lots are large, you're safe from the threat of a multistory building rising suddenly on your south flank, and you have some leeway to position your collectors at a safe distance from neighboring shade hazards.

Second, examine the neighboring property on the south side of your house. If there are signs that the trees may grow high enough to cast shadows over your house, or if your house is close enough to the property line that it might be shaded should your neighbors add a second story, reconsider your plans to add a solar heater. It's of no use to incur the trouble and expense only to have the system rendered inoperable in 3 or 4 years because of shading. If you have another potential collector area, unthreatened by shading, on your property, place the collectors there. If not, you have the option of finding a new lot in a more favorable spot.

Some states have gone as far as to permit legal action to be taken to secure sun rights. In Colorado, for example, you can follow procedures for creating and recording voluntary solar easements—legal agreements with your neighbors that guarantee your right to direct south sun (just as you can guarantee your access to a beach in a seaside town).

In Oregon, planning departments are attempting to take solar considerations into account in land use planning and zoning requirements, allowing planning commissioners to recommend ordinances that would govern building height to protect solar heaters.

In New Mexico, legislation has gone one step further: sun rights are guaranteed on a first-come, first-served basis to the owner of any solar collector producing more than 25,000 Btus of heat on a clear late-December day.

But it may be a while before the question of sun rights is settled on a nationwide scale. Until then, a homeowner installing a solar system does run a risk that the system may be shaded at some point in the future by a change in the configuration of the neighboring skyline.

Reflection problems. Another legal difficulty that may confront owners of solar systems concerns the reflection that can be caused by the glass covers on a set of collectors. At certain times of the day or year, glass-covered collectors which are mounted at an angle may reflect painfully bright sun either into a neighbor's house or yard, or, more hazardously, into the eyes of drivers on the street. The homeowner may not be able to change the angle of the collectors without lowering the performance of the solar system.

If complaints are lodged against the homeowner, or if he runs afoul of the building department when the system is proposed, there is little he can do—except to replace the flat glass covers with ones made of a less reflective material, such as textured glass or a tough, heat-resistant plastic (which may be less expensive than glass and is certainly less susceptible to vandalism). Or he can install vertical wall collectors instead of slanting ones on the roof (see page 38), as they are far less likely to cause bothersome reflections in the high summer sun.

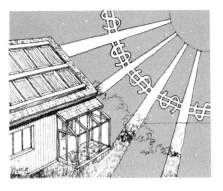

Will It Pay?
Solar Economics

For many people, the primary appeal of solar heating and cooling systems is that they run on vast, renewable resources of *free* sunlight instead of ever-costlier oil, gas, or electricity. Because of solar systems' independence of expensive fuels, it would be logical to expect them to be the cheapest form of home heating . . . but are they? Sunlight is free, but solar technology is not. In fact, most solar heating systems are much more expensive to install than systems that run on conventional power. The average active solar space heater can cost anywhere from $5,000 to $10,000, whereas an ordinary gas-fueled forced-air furnace will probably run in the range of $3,000. An electric water heater might cost $150 to $200; in contrast, a solar water heater averages $1,000 to $2,000.

(Passive solar systems are much harder to translate into dollars and cents than active ones; it's difficult to tell where house construction costs leave off and solar system costs begin. However, since most passive systems use only everyday building materials in their construction, and form an integral part of the house's structure, they tend to be less expensive than active systems.)

To compound the problem, almost no solar heater is intended to supply 100 percent of a house's heat. So an auxiliary heater must be installed (see "The Backup," page 31) along with the solar system, adding appreciably to the initial costs.

The high price of a solar installation may tempt you to ask, "Why bother to invest in solar if it's not economically competitive with a conventionally fueled heating system?"

The answer is that in many situations, solar *is* economically competitive with other types of comfort control, despite its high initial cost. The operating costs of conventional fuel-operated heaters and air conditioners are significant at the start, and they rise as the cost of fuel rises. A solar system's operating costs are negligible, however, because its primary fuel—the sun—comes to it at no cost. The secret of a solar system's economy lies not in its price tag, but in the money it saves after installation, over the course of its operating lifetime.

Life-cycle Costing

If you're concerned about the economic aspects of solar heating and cooling, you'll find that the most revealing way of calculating solar's cost competitiveness with conventional heating systems is through a method called "life-cycle costing." Instead of comparing only the initial component and installation costs of solar systems and conventional ones, the life-cycle costing method estimates the total costs each system will incur over its predicted lifetime—10, 15, or 20 years.

In order to perform a life-cycle costing analysis, you must include such predictable expenditures as the annual operating costs—electricity for the pumps, fans, and controls of an active solar system, or the cost of fuel necessary to run a conventional electric, gas, or oil appliance of the same heating capacity. (There are no significant operating costs for a passive solar system). Many analysts add an annual fuel inflation factor to the fuel cost figures in order to reflect predicted rises in fuel costs over the life-cycle period. Your state energy commission or local utility may be able to supply you with current predictions for fuel inflation; a common figure chosen for life-cycle costing is a 10 percent annual fuel inflation rate.

(Continued on next page)

Maintenance costs, such as replacement of mechanical parts with shorter lifetimes than the life cycle, replacement of antifreeze, corrosion inhibitors, polyethylene liners, or any other such equipment of an active or passive solar system, and cleaning of conventional and solar equipment, are also figured into the analysis.

If you plan to take out a loan or mortgage to pay the initial costs of a system (see below), you must add the annual interest payments and the initial loan fee to the total life-cycle costs as well.

Your analysis must allow for inflation rates, so that annual expenditures for operation and maintenance of a system will reflect the predicted 6 percent annual inflation over the chosen life cycle. And finally, you must subtract from your estimate of the initial costs any federal, state, or local tax credits, deductions, or exemptions available for solar heating installations (see below).

Your final life-cycle cost figures may surprise you. The initially higher-priced solar system may end up being the less expensive choice, once the repayment period on your loan ends and you've recouped the solar installation cost and the loan interest through your fuel savings, while the costs for fueling a conventional heater have meanwhile been soaring. (The longer your solar equipment lasts beyond the loan repayment period, the greater the savings you'll realize from your solar heater —so look for durability when you shop for your solar components.)

One example of a life-cycle cost comparison is a study done in San Diego, California, comparing an electric water heater with a solar hot-water system. The lifetime of the solar system was predicted at 20 years; its initial cost as $1,500. The electric heater's initial cost was $150, but its life expectancy was only 10 years, so a replacement cost of another $150 was added in, bringing the total to $300—still much cheaper than the cost of the solar equipment.

However, the difference between the two systems in fuel consumption

costs over the 20-year life cycle was phenomenal: $4,750 for the electricity to run the conventional heater versus $1,264 for the electricity to run the solar system's pumps and to provide backup heat during bad weather. In the final analysis, the solar hot-water system, at a total life-cycle cost of $2,689, was a much better deal than the electric heater, at a total cost of $5,050.

Most solar heating devices in most areas of the country pay for themselves in 10 to 20 years, depending on the device's performance and initial cost, local fuel costs, and the house's heating and cooling needs.

Giving You a Break: Taxes

In order to encourage energy-saving solar installations, the federal government and some state and local governments have proposed a variety of tax incentives in the form of credits, deductions, and exemptions. These tax aids can substantially reduce your initial installation costs, so you should keep yourself up to date on the tax breaks available for solar systems in your area.

On the federal level, the National Energy Act proposes a federal tax credit of 30 percent of the first $1,500 and 20 percent of the next $8,500 (allowing a maximum credit of $2,150), retroactive from April 1977 to the end of 1984. (At the time of this book's printing, this federal bill has not yet passed Congress.)

Various states offer similar credits and deductions from state income taxes, such as California's tax credit of up to 55 percent or $3,000 (whichever is less) of the cost of solar heating equipment and installation. In addition, many states allow a re-evaluation of property taxes for a house with a solar installation: some permit complete exemptions of solar equipment from property taxation; others allow solar systems to be re-assessed at less than their original value for property tax purposes. Legislation in some states exempts solar equipment from sales tax, and in others permits long-term loans in increased amounts for solar installations.

On the local level, still more incentives are provided, ranging from property tax exemptions to loan aid. In order to determine which cate-

gories of tax aid are available to you, write to your state legislator or Congressman, or write or call the National Solar Heating and Cooling Information Center (see "Solar Energy Associations and Information Centers," page 93).

One problem with most current tax incentive legislation is that it defines "solar system" quite narrowly: in general, only active systems bearing a certain percentage of the house's heating load will qualify. Passive systems are a long way from being standardized, whereas federal and industry-established performance standards for active solar equipment are approaching national acceptance. Since passive systems are so varied and their heat output is so difficult to quantify, the credit-worthiness of a passive solar system can be hard to prove to the tax board. However, you should check with your local tax-collecting office to see how the laws would apply to your installation.

In Pursuit of Mortgages and Loans

Life-cycle costing may make solar heating more economically appealing than its initial costs would suggest at first glance. But the high initial costs still deter many homeowners from embarking on a solar installation, simply because they can't afford to make the initial large investment of capital. For these homeowners, the dream of a solar-heated home can only be fulfilled with the help of a mortgage or loan to spread the initial cost over a period of years (as with the purchase of a house). Though available, these mortgages and loans are not easy to come by; many lenders are still nervous about the security of a solar heating investment.

Mortgage for a new house. If you're planning to build your house and solar system at the same time, you must obtain a mortgage in order to keep the interest at a reasonable rate. Sometimes burying the cost of a solar system in an 80 percent mortgage on a whole house can work to your advantage, because the solar part of the investment is less prominent. But difficulties can still arise.

The central problem in mortgage-hunting is that, due to the newness of the solar industry and the general lack of experience with solar, the property value of a solar home is difficult to determine. Since resale value is the crucial point on which a lender's willingness to grant a mortgage rests, many lenders are reluctant to "bet" on a solar installation. Lenders often comment that they think solar first costs are too high in relation to the system's market (resale) value. They also express concern over the reliability of solar heating equipment; they put little faith in warranties, claiming that few small companies can be relied upon to back them up.

Contributing to the difficulty solar heating investors encounter in searching for a mortgage is the traditional lending policy by which a borrower's income is compared to the cost of the house or system he plans to construct. Housing costs, including principal and interest payments on the loan, property taxes, and insurance premiums, should all total 25 percent or less of the borrower's income. Energy costs have only recently been considered as part of housing costs for this computation. So in the case of a solar-heated home, the lender often sees the principal cost as higher than that of a conventional home, but no corresponding discount is made for the predicted energy savings of the solar home. The homeowner's income may be considered too low for his proposed solar home, even though his energy savings in future years might offset the increased initial cost.

This lender uncertainty does not prevent solar homebuilders from obtaining mortgages at all, but it does make the task of finding a willing lender more difficult. Nor does it generally result in higher interest or shorter-term loans, since lenders feel that such a policy profits them little in a case of default. But lenders do tend to reduce their risks by imposing higher down payments and by undervaluing the solar system to allow for a smaller total loan than the system's actual cost might warrant.

Nevertheless, different lenders have different mortgage policies, and it is possible to arrange a mortgage with reasonable terms if you take the time to "shop around" and discuss your plans with a variety of lenders —mutual savings banks, savings and loan associations, life insurance companies—before settling on one.

Your search is more likely to be fruitful if you ask other solar home-owners and solar design firms to recommend lenders who have been willing to finance solar homes in the past. If you find no lenders with such a record in your area, try the larger banks, or seek out bankers who have a reputation for innovative lending policies.

You will also be in a better position if the performance of your system (or components) has been proven in previous installations. This will help lend credibility to your estimate of its value. (Some banks have a "good-product" list of solar equipment, developed by outside consultants, which they use to judge solar applications. Find out if such a list is available.) It is also important to bring your energy-savings calculations to the attention of the lender, to show how the savings will offset your high initial costs. And take heart—few moderately determined homeowners have found it impossible to obtain a mortgage for their solar installation, and many have found the process relatively easy.

Loans for a retrofit. Getting a loan for a solar retrofit is quite a different matter from finding a mortgage to build a solar home. A retrofit is considered a remodel, so the homeowner who wants to retrofit would apply for a home improvement loan—a shorter-term, higher-interest loan than a mortgage. This kind of application is easier and generally more successful, mainly because the important aspect of the application (in the eyes of the lender) is not the property value of the solar addition, which is hard to determine, but the homeowner's ability to cover the monthly payments—that is, the homeowner's credit rating.

The lender may wish to assure himself that the solar system is designed and installed by a reputable firm, but his decision on the loan will be based principally upon your credit-worthiness.

Refinancing—rewriting a permanent mortgage to cover the cost of the solar installation—is another option of a homeowner planning a retrofit, particularly if he has a long-standing good record of payment on his mortgage.

If you're looking for a loan on a solar retrofit, follow the same advice already given to those seeking a mortgage: ask for references among people who have successfully negotiated loans for solar installations. Then shop around for a loan with the best terms.

Some federally funded loans and grants are available to private homeowners who have low incomes, or plan solar heating or cooling installations with certain specifications, such as unusually low cost. To find out whether you might qualify, call or write the National Solar Heating and Cooling Information Center (address on page 93).

Passive Solar Houses at Work

The house forms its own solar heater
—ideas suited to new and existing homes

Embodiment of simplicity: *This house is a classic example of passive direct-gain solar heating. Large south windows collect winter heat (deep overhang minimizes unwanted summer heat gain); thick adobe walls and brick floors store it. When needed, heat moves naturally throughout home's open interior. Two thermosiphoning collector panels on bank below house heat domestic water. Architect: David Wright.*

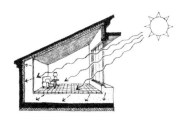
A Wall of Windows Faces South

Undaunted by snowy winters, this house relies on its windows, walls, and floors for heating, with only a woodstove for backup. For heat storage, the walls on the north, west, and east are concrete block (with few windows); the floors are also concrete. Insulation is 3-inch rigid-board on the exterior, covered with a stuccolike protective finish.

The whole south wall is double-pane glass windows to admit the sun for light and heat. (The painting studio is shielded from glare by the Trombe wall.) Insulating shutters, to be operated manually, are planned for the windows.

So much sun enters the house on winter days that overheating may threaten. To prevent it, a 1/2-horsepower fan in the basement draws warm air from the upstairs ceiling down through a duct and into a storage area under the basement slab, where the air gives off its heat before returning upstairs.

The storage area consists of concrete blocks, lying on their sides with their holes aligned to form long channels through which air can pass. If extra nighttime heat is needed, the fan circulates the house air again through the hot block storage.

The sun provides 70 percent of this home's heat—the average inside temperature fluctuates from 58° to 74°F. Just two cords of wood provide backup each winter.

Partial burying of the house helps stabilize winter and summer temperatures. Other summer cooling: an overhang to shield the south windows, operable skylights for venting, and good insulation and heat absorption by the walls and floor.

Dug into west-facing hillside, *this house is oriented to the south for solar heat. Left-hand window wall provides direct heat gain to living room. Heating for studio, sleeping loft comes from Trombe wall behind right-hand windows. Solar greenhouse at basement level on left provides heat to family room behind it. Windows swing open at top and bottom for summer ventilation. Architect: Stephen Lasar.*

Living room walls and floor *are exposed concrete for heat storage, softened by plants, weavings, and wooden furniture. Sun streams in through south windows at left; window in west wall opens view over valley.*

Hard-working Windows Admit the Sun

Temperature-sensitive "Skylid" insulating louvers, inside skylights revolve open when warmed by sun, allowing sunlight to enter. When sun is gone, louvers close, sealing in heat. (Louvers are tied shut in summer, to prevent overheating.)

Skylights or heaters? Windows set in south-facing roof planes of house's modules play both roles: they admit sun to light and heat rooms below. Panel on far left is one of two collectors for water heater. Redwood siding conceals massive heat-storing walls, floors. Designer: Zomeworks, Inc.

Rising gracefully out of a hillside, this house uses clerestory windows to pull sunlight deep into back rooms, buried underground. Front rooms bathe in light, warmth from sun entering lower rank of windows. Adobe walls dividing rooms provide extra heat storage as morning sun strikes their east faces, afternoon sun their west sides. Architect: David Wright.

Like steps of a staircase, this house lifts its windows to the low winter sun, each level absorbing and storing radiant heat in its own walls and floors. In summer, external wooden louvers shade windows. Architect: David Wright.

Sun-capturing windows are far from being a design straitjacket. They can take many forms, each bringing a different atmosphere and look to the house it heats. And they can be adapted to a variety of architectural styles.

Nevertheless, there are some rules in the design of direct-gain systems. The first is, of course, that most of the windows in the house must face south. The second relates to heat storage—the structure of the house itself. The windows' primary purpose is to direct incoming sunlight to the structural areas intended for heat storage—the floors, walls, and so on. So skylights, for example, are a good idea where the floor and rear walls are the predominant heat storage areas (as in the house at upper left, with its dark tile floors).

If a house is relatively shallow along its north-south axis, an unbroken wall of windows (like those shown on pages 48–49) works well because it allows the wintertime sun to strike all the way across the width of the house and up the rear north wall.

But if a house is deep along its north-south axis, the answer may be clerestories (like those at lower left) or a stepped-up series of windows (like those at right) to reach up and draw the sun down into the rear portion of the house.

Two other rules: Don't forget insulation for the windows to prevent heat loss. And summer shading is a crucial part of direct-gain design, to *prevent* heat gain when it's undesirable.

Activities match heat stratification: cooler bottom level (visible at right) is studio/workroom, where action keeps occupants warm. Middle level featured is kitchen/living room with fireplace. Warmest top level (not shown) is bathroom, sleeping area. Water-filled drums are buried in adobe walls that divide each level, providing extra heat storage. Windows offer extensive view as well as heat.

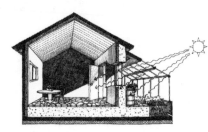

Greenhouses ... described on page 18.

Hybrid Solarium Resembles Greenhouse Heater

Here's a solarium that combines elements of a passive solar greenhouse and an active air system. It's attached to the south wall of the house, and its own south wall is almost all double-pane glass for heat collection, just like a solar greenhouse.

But instead of the massive, heat-storing walls of the usual solar greenhouse, this solarium uses an active system to store heat. A thermostatically controlled fan pulls the hottest air from the top of the solarium through a duct to an insulated rock storage area under the solarium's slab. The air gives off its heat to the rocks before returning to the solarium. When the house needs heat, the blower of the auxiliary oil furnace comes on and pulls heat out of the rock storage.

On winter nights, the solarium gets quite cold; the adjoining house walls are heavily insulated, and the doors and windows are double-glazed to prevent nighttime heat loss from house to solarium.

In summer, shading is provided by the wooden sunscreen attached to the solarium's south wall. Also, accumulated heat in the solarium may be drawn through a high vent by an attic fan and blown outdoors.

Household water is heated by five active water collectors mounted on the solarium's south-facing roof. Much of the year, these provide more heat than necessary for water heating alone. So when extra heat is available, the water from the collectors cycles through a heat exchanger in the oil furnace; the furnace blows air over the hot coils to boost the house's heat.

Selective sunscreen

Active solar water-heater collectors

Solarium

Two-story solarium occupies most of house's south side. Living room's sliding doors, bedroom's windows open into it, allowing sun-warmed air to travel directly into house. Sunscreen of wood on upper south wall of solarium admits low winter sun, screens out high summer sun (yet supplies indirect summer light, view). Active panels on solarium roof heat household water. Architect: Donald Watson.

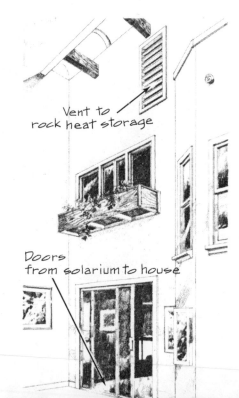

Vent to rock heat storage

Doors from solarium to house

Controlled heat flow from solarium to house is achieved using vents, fans, ducts: solarium itself has no storage mass except its concrete floor. In winter, fan pulls hot air from top of solarium down to rock storage under slab. Heat from storage is used in house when needed. In summer, solarium heat is vented out through roof.

The Center Is a Solar Greenhouse

During its frosty winters, this house derives most of its heat from its greenhouse. Heat delivery takes place in three ways. The first is the simplest: the doors of the house's living space open to the greenhouse by day, letting heat flow in directly. At night, the doors are closed to prevent return of heat to the cooling greenhouse.

Second, the foot-thick adobe house walls facing into the greenhouse absorb heat during the day. At night, the stored heat radiates slowly into the rooms.

The third method is active: fans pull the hottest air from the top of the greenhouse down through ducts to two insulated rock storage beds, one under each wing of the living area. Heat simply radiates up from the beds through the tiled concrete floors. (The beds can store enough heat to warm the house through two sunless winter days and nights.)

In summer, high and low greenhouse vents are opened to let hot air escape and pull cooler air in. The house walls themselves are shaded from the high summer sun by a triangle of roof.

Other design factors help the house hold heat: extra insulation in the outer walls, and a north wall bermed up to 4 feet to give less wind resistance and retard temperature fluctuations inside. The window in the breakfast nook faces southeast to invite the sun's first warm rays in the morning, while the living room looks southwest for late afternoon sun. Household water is heated by an active collector.

In midwinter, plants flourish in warm greenhouse environment. Some heat is absorbed by stone floor, sunken planting beds. Adobe house walls soak up more, radiate it into living areas at night. Some daytime heat flows directly into rooms through doors open to greenhouse.

A warm embrace: V-shaped house wraps around its lofty solar greenhouse. Two stories high, greenhouse's double-glazed south face lets sun flood in, traps heat. Hot air rising to top of greenhouse is ducted to rock bins under northeast, northwest house floors for nighttime radiation to rest of house. Architect: William Lumpkins; solar designers: Wayne and Susan Nichols.

Solar Greenhouses Double as Patios or Entryways

Not all solar greenhouses are designed to fulfill the bulk of a house's heating needs. Some are smaller, planned to provide a daytime heat supplement to the house and to act as pleasant, plant-filled extensions to the rooms to which they are attached.

The two greenhouses shown on this page have little or no storage mass in which to hold heat for carryover—certainly not enough to carry a whole house through a freezing night. The greenhouse/patio at right stores enough heat in its tile-and-concrete floor to keep its plants alive on severely frosty nights. The entryway greenhouse was built for a mild climate where danger of freezing is minimal; it has no storage mass at all. Yet both greenhouses supply a substantial proportion of their houses' daytime heating needs.

Solar greenhouse/patio provides ample winter daytime heat while active system on roof stores heat away for nighttime use. Buried in earth to waist-height for added insulation, greenhouse is covered with two layers of heavy polyethylene (stretched taut to prevent wind damage) to admit sun, trap heat. In summer, polyethylene is removed completely; area becomes outdoor patio. Designers: Wayne and Susan Nichols.

Inside greenhouse/patio are lush plants, warmth, and winter humidity given off by plants, hot tub: all are most welcome in dry, frosty climate. Doors to house are opened during day to allow heat to pass inside, closed at night. In summer, breezes cool open patio while permanent overhang shades house wall.

Front entry is solar greenhouse, supplying daytime heat and a green oasis to this owner-built home. Greenhouse is sealed off from living room at night by heavy sliding door to keep household heat from escaping through greenhouse's abundant south glass. Designers: Richard Jones and Steve Gilmore.

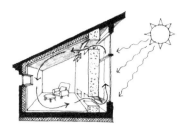

Addition Features Trombe Wall with Insulating Curtain

When the owner of this house decided on a major addition, he had it designed for solar heat, since his region's climate is ideal for solar— with cold winters but crisp, clear skies.

Accordingly, the addition was oriented to face south and built to conserve heat. And its south wall was designed as a Trombe wall solar heater.

The Trombe extends from the roofline down below floor level. It works like a typical Trombe (see page 17) during the day.

At night, when the room begins to cool, a sun-controlled motor lowers a multilayered, aluminized fabric curtain between the wall and its outside glazing. The curtain gives double protection against heat loss: it traps warm air between its many layers, inflating to fill the space tightly and insulate the wall; it also reradiates escaping heat back to the wall's surface.

The wall's configuration prevents reverse convection at night: while the curtain closes off the top vents to arrest escaping hot air, cool air settles to the bottom of the wall, below room level, and cannot rise into the room.

Inside, Trombe wall's inlay of rough-hewn fieldstone provides extra storage mass as well as textural interest. Stained-glass centerpiece window serves dual purpose: permitting light and heat to enter and giving access to insulating curtain. Rectangular openings at wall's top allow hot air to rise into room during day. Lowered curtain seals openings at night.

Behind two layers of glazing rises a 16-inch-thick Trombe wall, designed to heat new addition to house. At night, "curtain" of many layers of aluminized fabric is lowered between wall and glazing, pressing tightly against glazing to insulate wall. Architect: David Finholm; solar consultant: Ron Shore.

Trombe Wall House on a Small Town Lot

All rooms in house line up along Trombe wall for heat gain. Downstairs has concrete floor for extra heat storage, fireplace/heat-recovery device. Convection heating vents in Trombe are located near ceiling and floor (see above top).
  Light plastic dampers attach over top vents on room side; they float inward when convection carries hot air into room, seal shut when convection reverses, preventing escape of warm room air to cooler exterior. Arch in Trombe opens to greenhouse (above bottom)—welcome heat and humidity flow into house on winter days (curtain shuts off cooling greenhouse at night). Greenhouse heat storage is water-filled black steel drums, black-painted concrete floor, part of Trombe.

The challenge for this architect/homeowner was to build a passive solar home on a small lot in the midst of a small university town. To do this, he had to solve problems of lot configuration and position of shade trees. The solution was to apply for a zoning variance to permit placement of the house close to the lot line and sidewalk, in order to escape shadows cast by a neighbor's house and to save a venerable tree as summer shading for the house's west end.

The choice of solar system was decided by the Trombe wall's appealing simplicity—it combines collection, storage, and distribution in one south-facing, 15-inch-thick concrete wall. Its longevity (it will last the life of the building) makes it economical from a life-cycle standpoint, especially as it is combined with inexpensive overall house construction. Its low operating temperature makes it highly efficient, giving, as the homeowner pointed out, "high Btu per dollar investment." Finally, the Trombe wall is particularly well suited to the property's northern latitude and snowy winter climate, because with snow on the ground to act as a reflector, the vertical collector face is an optimal tilt angle for solar exposure (see page 38).

Aside from the cast-in-place Trombe wall itself, the house is standard wood frame construction, but with heavy insulation: R-18 in the walls and R-40 in the roof. Even the greenhouse was a standard commercial model, simply attached to the Trombe wall and outfitted for solar heating and cooling with double glazing, storage mass, vents, and shades.

The system has performed

well, keeping the house at an average daily winter temperature of 63°F. downstairs and 67°F. upstairs (though the evenness of a Trombe's radiant heat makes it seem warmer), with a nightly swing of about 5 to 10°. (In summer, average temperatures range about 10 to 15° higher.) A completely separate, thermostatically controlled gas furnace comes on when the temperature drops below 62°F.—still, throughout the entire winter of 1976–77, the house consumed only $75 worth of fuel, while neighboring houses of similar size gulped $500 worth of fuel.

The owner plans to install a door at the top of the stairs that can be closed to control heat migration up the stairwell, which has tended to rob the downstairs of some winter heat and overheat the upstairs in summer.

South face of house is two-story concrete Trombe wall, *covered with double glazing. Windows are cut into Trombe wall at regular intervals for light inside house. Attached greenhouse supplements Trombe heating, adds space for winter gardening in snowy climate. Summer house cooling: square eave vents at house peak open to vent hot air out from between Trombe wall and glazing, aided by tiny fans. Current of escaping air pulls hot air out of house; cool air enters through north windows. Greenhouse cooling: shades are lowered; windows at greenhouse peak open to let hot air out; vents in greenhouse foundation pull cool air in. Architect: Doug Kelbaugh.*

North side of house *has minimal windows, all triple glazed to cut heat loss. In winter some are fitted with fabric-covered insulating panels. Come summer, those located low in second-floor north wall are opened to allow cool air to enter house as hot air vents out top of Trombe wall.*

Visible behind glazing *is window embrasure in Trombe wall—flared to admit more light, glazed to prevent heat loss to outside glazing at night. White blind is lowered on summer days to reflect sun, shade room beyond. Rectangular vents high and low in Trombe permit daily convection heating loop (see page 17).*

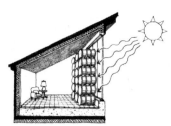

Water Walls ... described on page 17.

Passive Ways to Use Water for Heat Storage

The speed with which water can absorb heat, and the large amount it can hold in relation to its volume, make it very attractive as a heat storage medium. The main challenge in designing passive water heat storage is containing the water in such a way that it gets the necessary exposure to the sun.

Solutions to these problems of container choice and positioning in the house vary from system to system, as the houses pictured on these pages show. Containers are usually metal (for quick transfer of the sun's heat to the water inside), with the sides that face the sun painted a dark color: dark brown, green, blue, maroon, or black. Some designs use galvanized steel drums (though they

Heating columns become cooling columns when summer awning rolls down over collector windows during day to intercept sun, prevent heat gain in columns. Water in columns, cooled by nighttime venting through openings in south window area, absorbs daytime heat accumulated in house interior. The hotter the climate, the more water mass is necessary to keep house air cool.

Tall, dark water-filled columns of galvanized steel culvert line up behind a double-glazed, south-facing window wall to soak up heat from low winter sun, store it, and radiate it to interior as needed. Capped and bolted to foot-thick concrete slab floor, columns hold 1750 gallons of water, treated against rust and algae. House interior is open to encourage heat circulation; fan pulls heat into a few low rear rooms. Climate is mild; only wood stove augments water "wall" heating. Designers: Gregory Acker/Living Systems, Marshall B. Hunt.

are bulky); others involve containers made from lengths of steel culvert pipe or industrial ducting, welded to a base and capped to prevent evaporation, corrosion, and heat loss. The more south-facing surface area a container has per gallon of water it contains, the better solar exposure and faster heat gain the water will have. Containers can be lined with polyethylene to prevent leakage, or the water can be treated with rust and algae retardants.

As for placement in the house, the choice depends first on where the maximum amount of sun comes in. Pipes can be erected like columns (or drums stacked on their sides or on end) and ranked immediately in front of south windows. Or, if skylights or clerestories admit sunlight to strike north walls or angle back into north rooms, the water containers can take less conspicuous positions in the house, to avoid blocking south views and light.

In mild climates, summer shading may be the only protection necessary for the water storage (allowing passive cooling to take place—see the house at left).

However, in colder climates, movable exterior window insulation may be necessary to reduce winter heat drain, especially if the water containers are just inside the south windows. Also, more storage mass and a greater collection area (hence greater solar exposure) may be necessary for adequate heat storage. (If wall space is limited, greater solar collection can be achieved through the use of reflector-lined insulating panels—see page 38.)

Tucked against rear wall to free south windows for light, view, and direct heat gain, water-filled steel culvert columns look like part of house structure, but function as heat storage mass. Row of skylights directs low winter sun across room to strike and heat columns. To prevent heat loss on cold nights (or to exclude sun's heat in summer), wood-covered insulating panels are lowered along tracks in ceiling to cover skylights. (Insulating panels are manually operated, held in raised position by counterweights.) Designer: Michael Corbett.

Sculptural-looking storage mass is water-filled steel drums ranked on metal framework inside south-facing window, black ends facing sun for heat absorption. Crank on framework operates exterior insulation/reflector panel. Panel lifts up against window to insulate; cranks open to horizontal position to let sun in and reflect extra sunlight onto drums. Designer: Steve Baer.

Heat storage aloft: water-filled drums are stashed out of the way atop sturdy rafters. Rows of south-facing clerestories pop roofline up to scoop sunlight back into both north and south rooms, heating drums. Reflectors on roof bounce 30% more winter sun into house through clerestories. East windows angle southward for morning solar gain. Architect: William Mingenbach/The Architects, Taos.

A Wall Literally Filled with Water

In this adobe house, water for heat storage has been literally encased in a concrete wall, making the water storage a permanent part of the building. This system's advantage over a solid concrete wall is the speed with which the water picks up the sun's heat and transfers it (by conduction and convection) through the wall's mass to warm the room air. The concrete-and-water combination can also store about twice the heat that a solid concrete wall can.

Compared with other forms of passive water heat storage—drums and columnlike tubes—the water wall in this house takes up far less floor space and has minimal visual impact. Water walls of this kind do, however, involve more complex construction techniques than other types of water storage.

This house was designed so that the two water walls—one at each south end of the semicircular building—heat only the rooms they face; the central portion of the house is heated by a simple direct-gain system, supplemented by a central adobe fireplace.

Like other passive water storage containers, these walls can reverse roles from heating in winter to cooling in summer. Insulating panels are left closed during the day to prevent solar heat gain. Meanwhile, the enclosed water, cooled at night by venting to the outdoors, absorbs any excess heat in the rooms.

Heat storage wall is an elongated water tank—a thin-sided, precast concrete case, lined with polyethylene to deter leakage and filled with water. South surface is painted black, and double glazing covers wall to trap sun's heat. Reflector-lined insulating panel drops down on cold days to bounce added sunlight onto vertical wall. At night (and on hot summer days), panel is raised to cover wall area and hold heat in (or keep it out). Vents at either side of glazing release heat from wall to outdoors in summer. Visable at left is a similar water wall that heats house's other wing. Architect: William Lumpkins; solar design: Wayne and Susan Nichols.

Inside, wall looks like any other, but on sunny winter days, heat stored in water makes inner surface hot to touch. Crank mounted in room raises, lowers exterior insulating panel. Panel covers only wall area when closed; windows above wall (formed by upper portion of wall's exterior glazing) remain uncovered to admit light to room. (An overhang shades windows in summer—see drawing above.)

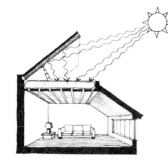

Roof Ponds ... described on page 19.

On Top of the Roof: Water Troughs

Built in a mild-to-hot climate with few freezing days in winter and very hot summers, this house has a roof pond system designed to fill cooling needs as great as its heating needs (if not greater). The system has proved successful: during the three hottest summer months, interior temperatures never rise above the 70s. Even on a blistering July day—112°F. outside—the interior was a comfortable 77°F.

How does the system perform such miracles? The secret is the insulating panels that cap the rooftop water pans. Each panel is wood-framed, topped with sheet aluminum to add strength and shed rain. The underside is 1-inch-thick rigid insulation, lined with a reflective surface; fiberglass insulation fills the panel interior.

By raising or lowering the panels over the roof ponds at strategic times, the homeowner can use the ponds either to heat or cool. In winter, the panels are opened daily so that the water can absorb the sun's heat. At night they're shut tightly so that the pans radiate heat to the house below instead of to the sky.

In summer, the lids open at night to let heat escape. During the day they remain shut to block sun from the pans, and the cooled water absorbs any heat rising from the house below.

The heat-conserving design of the house facilitates the roof ponds' operation: insulated window shutters supplement heavy insulation; the concrete floor slab and water-filled steel drums act as additional "heat sinks" to store heat for winter nights; north and south windows can be opened in hot weather to let cooling breezes pass through the house.

Sturdy roof joists and posts support heavy water-filled roof pans. Undersides of metal pans are exposed to house interior to facilitate heat transfer from water to house below. Black-painted steel drums behind couch, filled with water, act as extra heat storage, along with concrete slab floor. Direct heat gain comes from south windows. Hinged accordion insulating shutters unfold to cover windows, hold in heat at night.

Doffing cap to sun (or so it seems), house is in heating mode: insulated lids of roof pond heating system are raised, supported by motor-operated hydraulic jacks, to expose troughs of water to sun for heat gain. (Panels' shiny inner surfaces reflect extra sunlight down onto water pans.) Pans are set along length of roof peak, coated inside with black asphalt to absorb solar heat and curtail leaks, and each is filled 1 foot deep with water. Small skylights are set between pans. Designer: Jonathan Hammond.

Roof ponds

Insulating shutters

Concrete floor & steel drum heat storage

Reflective insulation panels over roof ponds

Roof over Roof Ponds in Wintry Clime

The new south-facing wing added to this farmhouse was designed to provide its own solar heat, using roof ponds. However, the site's far-northern latitude and the freezing temperatures and snowstorms common in winter there made some design adaptations necessary. The two most serious difficulties were the potential for freezing and the winter heat loss that would occur as heat reradiated from the ponds to the cold air, even while the ponds were collecting solar heat. Moreover, the wing was to be two stories high, so getting the solar heat down from the roof ponds to the first floor presented an additional problem.

The solution to the first two difficulties was to build a permanent, heavily insulated roof over the ponds, in place of movable insulating panels. The skylight, set into the roof's south face at a tilt of 60°, is the collection area. The skylight's double glazing traps the daily accumulation of solar heat in the ponds, and nighttime freezing is avoided as the ample insulation retains most of the day's heat in the attic.

Heat is distributed as follows: a blower pulls attic air, heated by the roof ponds, down through ducts into an insulated crawl space beneath the first floor. Heat then percolates up through five vents to the first floor, returning up a staircase to the attic. Additional hot air from the attic flows down another small duct to the second floor.

For summer cooling, a fan blows heat out of the attic to the night sky; window insulation remains in place during the day to prevent heat gain in the roof ponds. Cooled air from the attic is then blown through the house.

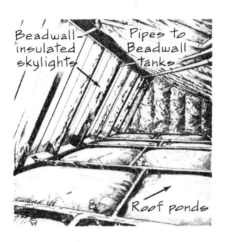

Like an attic full of waterbeds, roof ponds—actually black vinyl bags filled with water totalling 2200 gallons, nested in a wood framework lined with extra vinyl—are protected from harsh climate by insulated roof. Low winter sun enters south-facing skylight/Beadwall on left, strikes ponds, and bounces off foil-backed north-roof insulation for additional heating. At night (or during day in summer), beads are blown through pipes to fill space inside Beadwall.

Dominating south roof of new wing, wide skylight functions as both collector area and insulation for roof pond solar heating system located under protective roof. Skylight doubles as Beadwall: two layers of translucent fiberglass, with 4-inch space between, are blown full of polystyrene beads when insulation is called for, sucked empty when sunlight is needed to heat ponds (see page 12). Architect: Bruce Ellis; solar consultant: Bruce Anderson.

Buried in a Meadow: A Direct-gain House

The site is an unprotected field; the climate is cool and foggy. Ocean breezes and storms are frequent. The view seaward is to the west, yet the winter sun, as always, hovers to the south.

The design of this passive direct-gain house comes to terms with each of these factors successfully. The house is buried underground and surrounded with berms to deflect the heat-stealing winds and retard temperature fluctuations, both winter and summer. Only the view tower thrusts above ground to look westward, its lower level still snug below ground.

Underground construction must be solid to avoid heat loss and water damage and to support the earth. This house has 2 by 12-inch roof joists; the walls, floors, and roof are fortified on the outside with layered moisture barriers and rigid insulation. Elaborate drainage was provided behind walls and in the courtyard.

The house was designed compactly, with low ceilings. Backup is a single woodstove, but the house is so efficient that even on foggy days, when only diffuse sunlight filters through the south windows, the rooms are cozy.

Light-filled interior gives no hint of underground gloom; sun strikes clear across shallow, 20-foot-wide rooms. Floors of brick-on-sand, walls of foot-thick concrete store heat for continued comfort after dark. Window insulating shades—layers of aluminized polyester quilted between layers of cotton to trap air—slide up along standard sail tracks to let sun in, down at night to trap heat. On unusually hot fall days, shades stay down; lower windows and high vents in ceiling at windows' tops are opened. Air heated between shades and windows rises out ceiling vents, pulling cooling drafts through house after it.

South windows tilt steeply to let low winter sun penetrate deep into house, striking across floor to rear heat storage wall. All rooms (except bathroom and view tower) line up along double-glazed south window wall for direct heat gain. Thermosiphon collectors mounted in south windows at right heat household water. High berms surround sunken south garden courtyard to shelter house and garden from windy climate, help hold sun's heat.

Here's a house, not a hillside: west-facing view tower gives it away. Rest of house is buried under an insulating blanket of earth, held in place by natural meadow grasses. North skylights pop up out of roof slope, bringing light down into house's windowless north side. Skylights open when needed to vent excess heat. Architect: David Wright.

Active/Passive Hybrids Go Underground

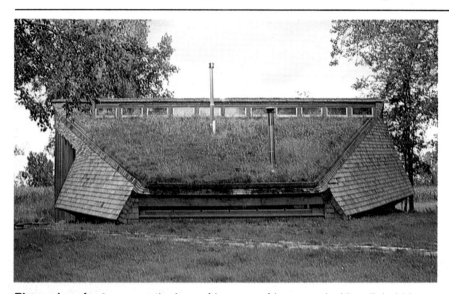

The sod roof returns: *north slope of house roof is covered with soil, held in place by carpet of grass, to help keep temperature stable inside house. House rises above high north berm to allow for row of low north ventilation windows, used in tandem with high north-facing clerestories for summer ventilation cooling. Winglike east and west walls, trapezoid shape of house are designed to offer minimum resistance to prevailing winds, thus cutting heat loss. Architect: Dennis R. Holloway and Univ. of Minn. Environmental Design Studio.*

Majority of windows *face south for winter heat gain (overhang gives summer shading). Solar greenhouse adds a share of heat, extends growing season to 11 months despite freezing winters. South roof angles at 60° to face winter sun directly, bearing a variation of the trickle-type collector, known as a "sandwich panel" collector, for active space heating. Berms on east, west sides of house come up to eaves; basement is completely underground.*

Underground construction and active solar heating devices make a happy combination, especially in climates where the direct sunlight needed for active solar heat collection is not an abundant resource. Though these two houses are built in very different climates, they share the problem of haze and overcast, making active solar collection an intermittent phenomenon at best, and requiring a doubly rigorous conservation of heat to keep the houses comfortable.

The climate of the house shown at right is relatively mild but damp; freezing temperatures are rare, but the rainfall is very high year-round and summers can be uncomfortably warm.

By sinking his house underground, the homeowner/architect was able to take advantage of a temperature equilibrium of about 52°F.—the average temperature of the local soil 4 feet below the surface. Despite the heat drain through the few exterior windows, the earthen insulation provides temperature stability that cuts the house's heating and cooling demand drastically, leaving less work for its solar-assisted heat pump (see page 32). Located on a suburban plot, shoulder to shoulder with the neighbors, the home's underground aspect also provides privacy and a welcome noise reduction.

In such a moist climate, special precautions were necessary in the house's construction to allow it to shed water easily and bear heavy loads. The concrete and heavy timbers used as structural supports were designed to carry up to 250 lbs. of water-saturated earth per square foot. A total of 3 inches of rigid-board insulation, topping a waterproof

In east view over house, *projecting collector areas are all that is visible, aside from rim of sunken atrium beneath south collector slope. Door at right leads to stairs, giving house direct access to roof. Roof area itself will be landscaped, becoming extension of garden. Skylights rising out of soil transport natural light deep into corners of house farthest from atrium windows. Architect: Norm Clark.*

membrane of five-ply hot-mop fiberglass and flashing, wrapped the house's sturdy roof deck before the soil covering (actually 6 inches of crushed gravel topped by straw and a foot of soil) was heaped over the house. In addition, perforated tile was laid underground, all around the house's foundation and wherever the roof met the supports, to carry water away from the structure and prevent leakage.

The house at left is built in a region of severely cold, gray winters and hot, relatively humid summers. Consequently its partially underground, sod-roof design is an important factor in reducing both its heating and cooling loads so that the active liquid trickle-type system can fulfill a high percentage of the house's heating needs.

The sod roof and high east, north, and west berms front the worst winter winds, acting as a wind barrier as well as insulation. In summer they aid in cooling, not only by insulating the house, but also through transpiration of the grass growing on the north berm. High north-facing clerestories, up at the level of the house's sleeping loft inside, open to let heat escape from the roof peak. Meanwhile down below, the low north vents open to allow a draft of air to enter the house through the carpet of grass growing on the north berm immediately outside the vents. The air is thus cooled by transpiration before it enters the house vents.

On an ordinary suburban site: *an unconventional house, dug into an eastern hillside and covered over with earth. An earth berm wraps around house's eastern facade, submerging house almost completely, save for row of east windows that welcome rising sun's first warmth into kitchen and dining room in morning, and soaring slopes of south-facing roof where collectors will be mounted to assist electric heat pump with space heating.*

Wrapped around a central atrium, *underground house opens each room to courtyard for natural light through glass doors and windows. Winter sun strikes deep into south-facing living room for passive heat gain; overhang screens out summer sun. Sand-set bricks over gravel pave atrium, raised in center for drainage into peripheral flower beds. Beneath atrium, perforated pipe gathers runoff, funnels it through pipe under house and away from foundation.*

PASSIVE/UNDERGROUND **65**

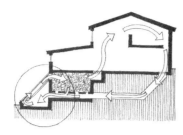

Natural Convection Heater: No Moving Parts

Though the heating system for this house resembles an active solar heater—with its collector panel area, absorber plate, ducts, and rock bin—it can nevertheless be classified as passive, because it relies solely on natural convection for its heat distribution. No stage in the solar heating cycle requires any moving part or input of conventional energy. There are no pumps, fans, or thermostats: what controls there are— dampers and windows—are manually operated.

The heating cycle begins when the sun heats the air between the collector glazing and the layered mesh absorber plate. Through its own natural properties, the hot air rises, unimpeded, directly into a rock storage bin situated in the house's basement, above and behind the collectors. When heat is needed in the house, dampers are opened manually in the house floor above the rock bin, and heat floats up through the vents and into the house. The open interior of the two-story house encourages the heat to circulate freely, rising naturally into the second-story sleeping lofts. Then, as the air cools, convection currents tug it to return vents in the floor and down through ducts to the bottom of the collector once again for reheating.

In summer, the air collector vents to the outdoors to avoid overheating the house. But the household water, heated and transported up to the house by the same thermosiphoning method as the air, continues to heat and cycle into the house throughout the summer.

Extending down hillside below house like a shed, thermosiphoning air collectors (and domestic water heater) face winter sun. Position of collectors allows natural heat distribution to house above without mechanical aid. Behind collectors, but in front of rock storage bin below house, is space for narrow workroom. Row of southfacing windows on house's first floor permits direct-gain heating in winter; overhang offers summer shade. Solar engineer: Zomeworks.

A closer look reveals black PVC pipe at left, used in thermosiphoning collector that heats household water. Visible further to right, air collector's absorber plate consists of layers of black-painted expanded metal lath; air sifts through, picking up heat and rising to rock storage behind and slightly above collector area.

A Sampling of Passive Water Heaters

Passive water heaters and pre-heaters abound in many different and ingenious forms, of which a few examples are shown here. One system takes advantage of an existing Trombe wall space-heating system to preheat house-hold water at little added cost beyond the extra copper pipe. Another solves the problem of a gently sloped roof, yet avoids the ongoing cost of electric pumps, by installing a thermosiphoning water heater and its collectors in a chimney housing, with the tank mounted higher than the collectors.

Still a third system uses a series of six small, connected tanks to heat and reheat water to high temperatures, using only the ever-present municipal water pressure and natural convection to move the water from tank to tank. Since cold water enters the bottoms of only the first three tanks, and hot water is drawn off the top of the last three, cold and hot never mix directly as water is drawn off and replenished, so the hot water stays hot.

Low cost makes this system even more attractive: the tanks themselves are simply liners for recreational-vehicle water tanks, and the glazing that covers them is formed by double-glazed patio doors. (In freezing climates, insulating shutters for the bay windows would be required.)

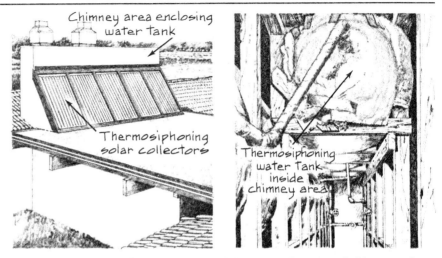

A tank in a chimney? Here, a thermosiphon water heater and chimneys share enlarged enclosure atop house. Collectors are securely mounted on outside south face of structure. As cold water from city main is warmed in collectors by sun, it rises naturally into tank. Climate is mild, presenting little freezing danger. In colder regions, heat exchanger and antifreeze must be used (see page 28). Designer: David Springer.

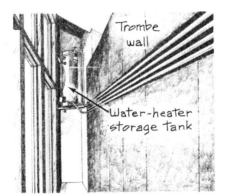

Trombe wall preheater: Cold household water passes four times across Trombe's south face through black copper pipe, mounted in parallel lengths in recessed area of wall. Besides receiving direct sun through south glazing, water shares Trombe's stored heat before entering uninsulated black tank mounted at west end of Trombe. Tank also absorbs sun's heat directly through south glazing. Electric coil in separate tank completes water heating. Architect: Stephen Lasar.

Six small tanks in two south-facing bays make up low-cost passive water heater. Each bay contains three uninsulated, black-painted tanks under double-glazing, angled south for adequate solar exposure in both winter and summer. Municipal water pressure moves water into first set of tanks for preheating; no pumps are used. As hot water is drawn off in house, warmest water at top of first bay's tanks moves to bottom of second bay's tanks to continue heating. Finally, hottest water from top of second bay's tanks moves through 20-gallon backup water heater in garage and out tap in house. Designers: Gregory Acker/Living Systems, Marshall B. Hunt.

Passive Systems Adapt to Existing Houses

Retrofitting a house with a passive solar heating system usually involves major remodeling, if the system is expected to provide a substantial portion of the house's heat. In addition to heat conservation precautions (adding insulation, weatherstripping, and double glazing), structural changes are often necessary. Two relative exceptions to this rule are the solar greenhouse and the Clearview collector. Both of these can be added to quite a variety of house types without major structural overhauling.

The greenhouse is perhaps the most adaptable, as long as the house offers an adequate expanse of unshaded south, east, or west wall (if it has windows and doors in it, so much the better) and enough room on the property to extend the house several feet without violating zoning setback restrictions. A greenhouse can be designed to house its own heat storage mass and provide its own summer cooling, and it can be added to any kind of wall.

The Clearview collector should be added to a masonry wall (temperatures rise too high for the wood in a stud wall) containing fairly high-set windows, but it has the advantage of requiring very little extra space on the property. It offers a low-cost, effective way to retrofit a house for passive solar heat. If the system is expected to provide overnight heating for the house, storage mass, such as water-filled drums, must be added to the house's interior to absorb heat during the day and radiate it at night. (The Clearview collector can also be used in new houses with or without fans and additional rock storage mass.)

Solar greenhouse is one possibility: This one was added to south wall of house in place of flower garden. Glass south windows and fiberglass roof admit low winter sun; overhang screens noon sun from greenhouse in summer. Blue water-filled drums (against house wall), brick floor, and existing brick house wall form heat storage mass for greenhouse; woodstove provides backup heat when necessary to keep plants from freezing. Greenhouse donates substantial daytime heat to house during much of winter through house's east and west doors, south windows. Cable on pulleys opens heavy insulating shutter high in greenhouse's north wall, revealing north-facing vents that release unwanted summer heat from roof peak. Architect: William Lumpkins.

North vent for summer cooling

Door admits heat to house

Water-filled drums, brick floor & house wall store heat

Vent to release summer heat

Clearview collector

Existing south wall becomes Clearview solar heat collector. To convert masonry wall to daytime solar heater, exterior surface was painted brown; ventilation holes were cut low in wall (operable window took place of holes cut high in wall). Double glazing was added to cover wall's entire exterior from floor to roof; dark Venetian blinds were hung between wall and glazing, with slats tilted slightly to face winter sun. Blinds absorb sunlight, heating surrounding air. Hot air rises into open high window; convection draws cooler air through low vents to absorb heat from blinds, rise, and return to house. Window is closed, vents are covered at night to prevent heat loss. On summer days, high side vents are opened to allow air heated in collector to escape, pulling hot air out of house with it. Designer: John Peck.

Solar Overheating?
Selective Screens Prevent It

In passive solar design, screening the sun out of the house during the hot part of the year is just as important as admitting it in the colder months. If no provision is made to stop the sun from flooding into the house all year around, winter solar heating will become summer overheating.

Some common techniques for screening the sun out are discussed on page 11 and page 12; others are included in descriptions of the cooling modes of various passive system types (starting on page 16). Shown here are examples of both shading to prevent overheating of a direct-gain system, and insulated sunscreening to permit a night-cooled water wall system to absorb the house's heat, rather than the sun's, during the day.

To store "coolth" for daytime cooling, skylights are left open at night, covered only with light insect screen, to vent any heat buildup from water columns, room air. Then, as morning sun first rises, chain-operated jack lowers insulating shutter down to fit tightly over skylights, preventing entry of sunlight or heat. Cooled water columns absorb any heat accumulating in room air during course of day. Designer: Jonathan Hammond and Jim Plumb/ Living Systems.

Jack & chain operate, support skylight's reflective insulation panel.

Trellis shading

Vine shading

Venetian blind shading

A cooling shade is cast over south-facing window area by a combination of screening techniques, preventing heat buildup in concrete floor slab on hot summer days. White Venetian blinds in windows can be turned so slats make solid white screen to reflect sunlight back out window. Trellis over deck area is covered with grass matting for summer shade. (Matting is removed in winter.) With windows at southwest corner of house, late afternoon heat was a problem; deciduous vine with dense foliage was grown over west end of trellis to block summer afternoon sun. Designers: Gregory Acker/Living Systems, Marshall B. Hunt.

Convection: A Key to Natural Cooling

Convection cooling is very familiar to those people who open all the doors and windows on a hot summer's evening and then switch on the big old attic fan and let it blow the day's heat out of the house, pulling sweet drafts of cool evening air in behind it.

These houses follow the same principle, but they use the sun's heat, instead of a fan, to start the convection current. They simply encourage the sun-heated air to follow its natural inclination to rise out high windows, and then give cooler air the opportunity to enter through low vents and take the hot air's place in the house.

Some houses rely on this cooling method alone, with a little judicious shading. Others go to greater lengths: encouraging prevailing winds to enter the house, or even overheating (and then venting) the highest part of the house in order to strengthen the convection current (and hence its cooling effect).

Still more complex variations exist: some houses use convection to draw outside air deep into buried "cooling tubes"—lengths of PVC pipe that travel underground for several yards before emerging inside the house. The cool "steady state" of the earth around the tubes chills the air on its way to the house's interior. Other variations pull air past moistened wicks to produce an evaporative cooling effect.

Convection cooling works best in dry climates, but even in humid ones the moving air offers some relief.

Cooling plenum, high above living space, uses solar energy to initiate convection current for cooling in hot, dry climate. Plenum is dark-painted metal box set into west wall of third-floor mechanical room. Inside box, channels contain rocks with air spaces between; duct connects box with house below. As plenum's exposed west face heats up in mid to late afternoon sun, rocks gather heat, pass it to air in box. To initiate cooling, low vents in house's first-floor walls are opened. Plenum-heated air rushes out wind turbine in roof; cooler air flows into house through low vents and up through duct to plenum. Hot storage rocks maintain air current into evening. Architect: Richard Crowther.

Raised floor, open foundation, and sets of high windows near roof peak team up with low interior vents to keep house cool. Warm air from rooms vents out open high windows, while vents near floor draw cool air in from house's perpetually shady, breezy crawl space. Deep verandas circle house to shade exterior walls. Architects: Gluth and Quigley.

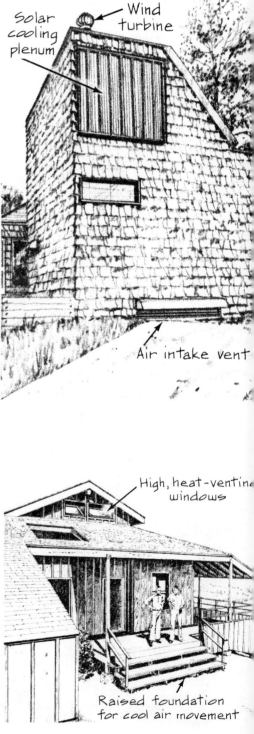

Solar cooling plenum

Wind turbine

Air intake vent

High, heat-venting windows

Cool air intake vent near floor

High, heat-venting windows

Raised foundation for cool air movement

No convection current more direct and simple: low north hatch on left opens to let cool air in; south windows above study loft at right open to let heat buildup escape. Refreshing drafts of air float up and across house's open interior. Designers: Richard Jones and Steve Gilmore.

House has disappearing walls to escape tropical heat: sliding wood panels and shoji screens slip aside, louvered windows and interior walls flip open to let trade winds waft through house. Shady, open house interior welcomes breezes into every corner. In cooler weather, everything shuts tight. Architects: Pearson and Terry.

Active Solar Houses at Work

Moving the sun's heat by mechanical means—adaptable ideas for new and existing houses

Trim and discreet, active system's collectors blend with roof of large, traditional Cape Cod-style house. Open-loop water system includes 2,000-gallon storage tank in basement, thermostat control, pipes, pumps, and fans for distribution, and forced-air backup heater. Collector covers are single layer of fiberglass, corrugated for extra strength; absorber plates are aluminum fins clamped to copper tubes, painted black. System drains down to prevent freezing. Sun provides around 50 percent of house's heat; south-facing window area adds supplementary passive heat gain in winter. Designer: Acorn Structures, Inc.

Active Hot Tub Heaters ... described on page 30.

The Sun Heats These Hot Tubs

Solar systems for hot-tub heating, like those for heating domestic water, require relatively small collector areas and can be quite compact in design, which makes them excellent retrofit options. Some are self-contained, as in the one below. Others have the collectors located on the roof at a greater distance from the tub. In fact, with some extra collector area, a solar system can provide all the summer heat and much of the winter heat necessary for both a hot tub and the household water, as in the system at right. The combination can be quite economical.

In mild climates, hot tubs are often located out in the garden and designed with handsome decks for outdoor entertaining. In these cases, a cover for the tub is important to conserve the solar heat while the tub is not in use.

In colder climates, solar-heated hot tubs are equally viable, but they must be located indoors for year-round use.

Eight panels on roof heat hot tub and household water. Homeowners added solar system and hot tub when they had roof rebuilt. Heated water from collectors goes first through underground pipes across garden to hot tub, then cycles to household water storage tank in basement. To heat tub water without danger of scalding, solar-heated water circulates through copper heat-exchanger tubes running around perimeter of hot tub's interior. Solar system heats tub to 103°F. on clear January days, much higher in summer months. Backup is gas heater. Designer: The Solar Center.

In a tiny city garden, this compact solar hot tub was custom-designed and owner-built. Tub heats up to 110°F. on a sunny day (and loses about 10° overnight). Shed supports one large 5 × 10-foot collector panel (finned copper absorber plate with one layer of fiberglass covering) angled to face sun. Shed also contains pump, thermostat, sauna, and solar-heated shower. Climate is mild, but during occasional freezes, pump circulates hot-tub water through panel. Designer: The Solar Center.

Solar Pool Heating: Flexible and Economical

Panel camouflage: *Unglazed black PVC collectors are mounted on steep black roof of colonial-style house to heat deep round pool below. Seventeen collectors were installed to keep pool warm (85°F.) all year around. Though most of the panels fit onto conveniently south-facing roof plane (divided into three clusters by chimney and vents), some were placed on east-facing roof plane, with less desirable solar exposure. Large pump circulates water from pool up to high roof. Gas auxiliary supplements solar heat on occasion. Designer: Fafco Solar Systems.*

Pond bank provides *unshaded, sloping site for solar pool-heating panels. Unglazed, black PVC panels lie directly on soil of south-facing bank; reflection off water adds to solar collection. Buried pipes lead uphill to terraced pool area off west end of house, carrying heated water to pool. Architect: Bruce Ellis.*

Collectors on the fence: *Two tiers of panels, mounted on retaining wall and deck rail that rings hillside pool area, face south and southwest for solar collection. Copper-absorber, single-glazed panels provide all pool heating (no backup for pool heat was installed). When pool is heated to selected temperature, collectors transfer extra heat to 700-gallon storage tank to preheat household water and radiant space-heating system. Climate is mild; panels heat pool from April to October. In winter, solar system is used only for household water and space heating; pool is left unheated. Architect: John Boyd.*

PVC panels on flat poolhouse roof are mounted on freestanding framework to get necessary angle and orientation for solar exposure. Pool is solar-heated from March to October, then gas auxiliary comes on. System can be controlled thermostatically or manually. Early problems with water circulation were solved by new thermostats, replacement of small pump with larger one. Designer: Fafco Solar Systems.

Compared with solar space heating, solar pool heating is remarkably adaptable. Its low operating temperatures and summer use free it somewhat from the stringent requirements of collector orientation and tilt. The relatively high summer sun—from May through September—is readily available, and a fairly low efficiency of solar collection is enough to provide all the heat a pool needs. (Of course, if you have other uses in mind for the collectors, such as heating a hot tub, household water, or even the house air itself, you need just as rigorous a design and as efficient a system as for any other kind of solar heating.)

As a result of this design flexibility, there are successful solar pool-heating systems in which the collectors crop up in all kinds of unusual situations: fences, poolhouse roofs, banks and hillsides, and garage roofs. Some even face directions other than directly south. (Beware, however, of sloppy design: just because there's room for more variation in solar pool heating design doesn't mean that the designer can afford to ignore the basic tenets of solar design in general.)

In addition to its suitability for retrofitting, solar pool heating has the advantage of economy: homeowners turn to it because it conserves resources or, as in one case in which the house and pool crown an open hilltop, because the sun is plentiful. Most often, homeowners turn to it because gas or electric pool heaters cost so much to run that they threaten to become unaffordable (or even outlawed in some states). A solar pool heating system, on the other hand, grows cheaper as it pays for itself.

Solar panels were installed on garage roof 1 year after pool was built, to coincide with roof remodel. Forward-looking homeowner/installers put in pipes for solar heating when they added pool, so as to conceal piping, make later solar installation easier. Five collectors (aluminum absorbers with copper tubes, single-layer fiberglass covers) provide heat from May to September, then gas heater supplements system through October. One more panel may be necessary to prolong swimming period and overcome shading problems caused by nearby pine trees. Built-in spa is gas-heated; temperatures needed are too high for this solar system to heat both pool and spa. Designer: S. M. Pace.

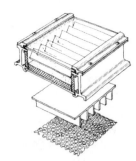

Air Systems ... described on page 24.

Wall of Hot Rocks Heats the House

The active air system that heats this house has an unusual storage arrangement: the masonry north wall serves as a well-insulated rock heat storage bin. Filled with 60 tons of fist-size rocks, the bin receives heated air from the collectors through ducts leading into its bottom. Hot air then travels up through the rocks, giving off heat as it goes. A plenum at the top of the storage wall can lead the air back to the collectors for reheating; if the storage is already hot, so that the air emerges still warm, it is blown to floor registers leading into the house. The ducts themselves are cleverly disguised as large beams to blend into the house's design of open cathedral-type ceilings.

Summer cooling is accomplished by blowing chill night air through the storage rocks to cool them. Then, during the day, fans blow the house air through the cooled rocks, where it gives off its heat.

The collector area itself was assembled on site and built right into the roof, which was designed at the optimal 45° tilt for winter sun exposure. The collector's absorber plate is steel, painted black and finned to disrupt the air as it flows behind the plate, improving heat transfer. A single glass layer covers the collector, and a 3/4-horsepower blower is the force behind the moving air.

Ceiling beams are really air ducts, carrying air between collectors on roof and vertical rock storage insulated within masonry wall at right. Storage provides a minimum of 2 days' heat carryover during winter storms. Heat-recovery device in fireplace provides added heat; gas furnace is main auxiliary heater.

Hot-air ducts from collectors to storage

Vertical rock storage in wall

Built-in air collector occupies most of the south roof slope of this conventional two-story, saltbox-type home. House itself is heavily insulated, with double-glazed windows arranged mostly on south side for passive heating assistance. Designer and solar consultant: Herb Wade.

Active air heat collectors

South windows for direct heat gain

Bold Solution to a Solar Design Dilemma

The homeowner/architect of this house tackled a tough design problem when he bought a plot of land on a steep, west-facing hillside, covered thickly with mature trees, on which to build a solar home. His solution was a narrow, towerlike house, rising four levels up from the slope for a broad expanse of south-facing wall space. Then he tilted the south wall to form a 460-square-foot collector area, cantilevered out from the house at a 60° angle to face the winter sun. The trees to the south of the house still cause some shading problems; the owners intend to top a few of the closest trees to improve the collectors' solar exposure.

The collectors are air-type, made of inexpensive corrugated aluminum roofing painted flat black and covered with double glazing formed by glass patio door seconds. Beneath the house lies the rock storage bin, poured in place along with the foundation and filled with 50 tons of fist-size river rock.

Thermostats and fans regulate the delivery of air from collectors to storage and from storage to house. In summer, the night-cooled rock storage is used to cool the house air.

The house's open interior, combined with the natural heat stratification produced by its height and many levels, helps the active solar system by allowing natural circulation of heat inside. The different levels are designated for activities to match their comfort level: an action-oriented family room is on the lowest level; the living room, with its woodstove, is on the next; above that, are the kitchen and dining room; and on the top and warmest floor are the bedrooms.

House's air collectors face south, though site's orientation is west due to steep-sloping hill. Collectors jut from south wall at appropriate tilt angle for solar collection. Above collector area, clerestories admit sunlight to top-floor study area. Trees at left were preserved for beauty but must be trimmed, topped to prevent shading of collectors. Architect: William Bishoprick.

Clerestory windows admit light

Active air heat collectors

Hot-air duct leads directly to house from top of collectors

Collector-to-house heating: hot air rises directly from top of collectors into house through duct leading into bathroom on house's top floor. Manually operated damper shuts off duct when heat is needed in storage or when collectors are cold.

Air Systems Adapt to Many House Styles

Though the ducts and rock storage bins of active air systems are space-consuming, they set no unusual limits on the architectural style of new homes built to incorporate them, beyond the basic requirements of southward orientation and energy conservation. Traditional-looking houses can shoulder their solar collectors as gracefully as "futuristic" ones, and their interiors can reflect an equally broad spectrum of life styles.

In addition to the concern many people show over the esthetics of solar systems, a frequent question is, "Are solar heaters troublesome?" Answers vary. Some owners of systems shown here have reported trouble-free operation, similar to that with a conventional heater. Others have had problems ranging from leakage of rainwater into the collectors to inaccurate sizing of rocks for storage to breakdowns of blowers and sensitive electronic controls. (Similar problems, in addition to incidents of corrosion and freezing, have been reported with active liquid systems.)

One homeowner/builder had to shovel all the stones out of his storage bin and replace them with larger stones. Another found it difficult to get prompt service from the manufacturers, even in an emergency (luckily, he was an engineer and could do some of his own servicing). Still others found that their manufacturers provided quick and helpful service and an equitable price adjustment.

These experiences are not uncommon among owners of new solar homes. They reflect the still-experimental nature of much of the present-day solar technology.

Deeply bermed mountain home bears 2/3 of its air-type collector vertically on south wall; 1/3 of collector is at 45° slant on roof. Vertical portion of collector gains extra winter insolation through reflection from snow on ground. Household water is preheated by passing through heat exchanger in duct leading from collectors to rock storage immediately behind and beneath collector. Architect: Richard Crowther.

Roofline of traditional Cape Cod-style house accommodates solar heating system well: air-type collectors on south roof tile a steep 45° to face winter sun. Heat storage bin—insulated plywood filled with 1/2-inch stones—is in crawl space beneath house. In summer, evaporative cooler is used during off-peak night hours to cool rock storage; storage cools house during day. Designer: Solaron/ Eco-Era, Inc.

Contemporary house has heat-conserving design *to ease burden on air-type solar heating system. Garage and berms buffer north and west walls; entries have double-door airlocks to retain heat when outer doors are opened. Windows are double glazed; south ones are deeply recessed for summer shade, west ones have reflective glass to minimize summer heat gain. Direct collector-to-house cycle of solar system is used on sunny winter days for heat, and on hot summer days for ventilation through collectors to outside air. Cooling plenums (see page 70) in west wall assist with ventilation. Architect: Richard Crowther.*

Colonial-style duplex *is heated by one active air system. Two halves of building share collector area, but in agreement with local building officials, each family has its own auxiliary heater, set of ductwork, and half of divided rock storage, so return air from each family does not flow into common storage area. Designer: Richard Splittgerber.*

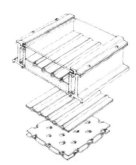

Liquid Systems ... described on page 25.

House on the Beach Uses Sun for Heat

Built on the beach, this house has an elegantly clean design that combines a number of active features with passive direct gain to make its solar heating system work exceptionally well. The basic solar heater is an active liquid type: 450 square feet of flat-plate collectors (all-copper tube-type absorbers with a selective surface and single layer of glazing); a closed-loop system using a mixture of water and non-toxic propylene glycol to prevent freeze damage; a 2,000-gallon steel storage tank in the basement that can carry the house through 48 hours of winter storm before the auxiliary oil furnace comes on.

The system has a collector-to-house cycle (unusual for a liquid system) which, when the storage is cold but the day is sunny, permits hot liquid from the collectors to circulate through a fan-coil device that blows the heat directly into the house. The liquid then returns to the collectors, bypassing the storage tank. Household water is preheated in a coil immersed in the storage tank.

To avoid winter daytime overheating from direct gain through the south windows, a fan and duct were installed in an interior living room wall. They draw heat from the south rooms and disperse it to the north ones.

Hot summers are usually accompanied by cooling breezes off the ocean, but additional ventilation is provided by two rows of north-facing clerestory windows backing the two rear collector mountings. These can be opened to vent excess heat from the house ceilings.

Aloft like sails, three ranks of liquid-type collectors top this beachside house; they stand apart on flat roof so as not to shade one another from low winter sun. Collector mountings are integrated with house structure for strength to withstand winds and storms. House is turned 10° west of south for ocean view, but still has good solar exposure. Broad planes of south window glass admit sun for daytime passive heating in winter; deep overhangs shade them in summer. Architect: Donald Watson.

Pleasant south living room gives no marked suggestion of solar design, but subtle features give it away: room is awash with light from expansive south windows; unobstructed view from living room up through study to master bedroom on upper level indicates open plan of house interior, designed to encourage free circulation of passively and actively gathered solar heat. Vents of heat-recovery device in fireplace are visible below mantel.

A (Nearly) 100% Solar-heated House

Dramatically successful in achieving its goal of almost 100 percent active solar heating, this house was built more as a demonstration of an active heater's capabilities than as an example of a practical solar application. Such a system is uneconomical not only because of its overall cost, but also because much of the collector and storage area is utilized only a fraction of the year, from November through April. The rest of the year it lies fallow, its huge quantities of heat unneeded in the house.

The solar system includes a full 1200 square feet of collector area, equipped with a drain-down system to prevent freezing, since the climate is very cold in winter. Water mixed with corrosion inhibitors is the heat-transfer fluid; it feeds directly into the 16,000-gallon storage tank which occupies two-fifths of the house's full basement (poured in place along with the foundation, then lined with insulation and water-proofed).

When heat is needed in the house, storage water is pumped through baseboard radiators in each room (the house has north and south temperature zones, each served by its own pump and circulation loop). Domestic water is preheated by circulation through a coil immersed in the big storage tank.

The solar space heating is so comprehensive that almost no backup is required. An ordinary 80-gallon electric water heater switches on when necessary to warm the storage water on its way to the radiators. But even in the record cold of the 1976–77 winter, the house was 92 percent solar heated.

North side of house, seen from street, gives no indication of comprehensive solar system that provides 90–100% of its heat. House's traditional design— well-insulated wood-frame structure with vinyl clapboard-like siding, dormers in roof, and ample north window space (albeit double glazed)—shows active solar's ability to adapt to accustomed life styles. Designer: Mark Hyman, Jr.

From garden, house's solar collectors are seen to occupy almost whole expanse of south roof, sloping at a 50° angle to meet winter sun. Absorber plates are aluminum tube-in-plate, painted flat black. Glazing is single sheet of acrylic. Collectors were assembled on site to fit roof precisely; those in upper row are slightly smaller than lower ones.

V-shaped Roof Forms Collector and Reflector

Though the style of this house is hardly conventional, its dual solar heating-and-cooling system is ingenious and functional. The system is a hybrid, using both active and passive techniques to keep the house comfortable.

For heating, the house's primary system is an active one: single-glazed collectors (aluminum tube-in-plate absorbers, coated with a selective surface) are built into the south-facing roof slope of the house's north half. They heat untreated water, which circulates in an open-loop system down to two concrete septic tanks serving as heat storage. The tanks are well insulated and waterproofed with an elastomeric compound; they keep the hottest water, which is used in the fan-coil distribution device, isolated from the less hot. A copper coil immersed in the hottest part of the hottest tank preheats the domestic water.

In summer, the active collector area converts handily to a passive solar-assisted ventilating system: the sharply angled, north-facing roof on the house's southern half is lined with reflectors that beam additional high summer sun at the collector area, heating it to very high temperatures. Vents leading from the house to a 6-inch air space inside the collectors (between the absorber plates and the insulation) are opened, as well as the house windows and a long vent running the width of the collector area at its uppermost peak. The intense heat of the absorber plates heats the air space behind them, and the hot air rises out the collectors' top vent, creating a convection current to pull hot air up and out of the house and to draw cool air in the windows.

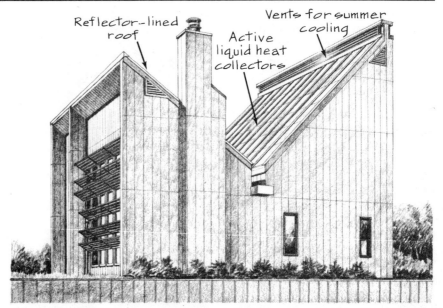

Dramatic profile of house serves solar heating function. Roofline on north portion of house slants up to face south at 45° angle, bearing liquid-type collectors. South portion of roofline not only veers upward to serve as reflector for collector area on facing roof, but also creates large south-facing wall for passive direct-gain heating. Unusual height of south window area required layers of overhangs, designed to admit sun in winter and screen it out—while still allowing view and indirect lighting—in summer. Designer: University of Tennessee.

Winter sunlight floods through double-glazed south window wall, heating living and dining rooms directly. Blinds and insulating shutters are closed at night to trap heat inside. In summer, exterior overhangs give shade; windows open to assist in ventilation of house interior.

Multi-unit Solar Housing Works Well

Solar heating is by no means limited to single-family homes. It adapts easily to multifamily housing, as long as there is the necessary south orientation and enough roof area to hold an adequate number of collectors.

In fact, from a solar viewpoint, adjoining multifamily units have two advantages over separate homes: they tend to be smaller, demanding less heat, and they share at least one wall (and often two) with neighbors, so fewer walls lose heat to the outdoors.

Space heating is more space-demanding than domestic water heating, making it a trickier design problem for multiunit housing, but it is becoming more frequently possible for a prospective solar homeowner to locate townhouses, duplexes, and even condominiums that have some form of solar heat.

Liquid systems are particularly adaptable to small-unit heating, since their pumps, pipes, and tanks take up considerably less room than an air system's bulky components.

The evacuated-tube collector shown at top is compact compared to a flat-plate collector (though it is also more expensive): each narrow tube-within-a-tube is a very efficient collector in itself. The inner tube has a selective black coating and acts as the absorber; the heat-transfer liquid flows through this inner tube. The outer one is glass, forming the "collector glazing." Between the two is a vacuum which admits solar radiation but cuts down on heat losses from the absorber to the outside air. The underside of the glazing tube is coated a shiny silver to concentrate even more sunlight onto the absorber tube.

Each of four adjacent townhouses has its own active liquid-type solar heating system, equipped with evacuated-tube collector array on sloping south roof and, in cellar, steel storage tank, pumps, heat pump, and backup electric water heater. Mini-computer controls complex system's operations: phasing in solar heating from storage to house first; if storage is not hot enough for direct heating, directing heat pump to heat house, assisted by warm storage; or, if storage is really cold, switching on backup to provide enough heat to heat pump. Heat pump also provides summer cooling. Solar designer: Ecosol, Ltd.

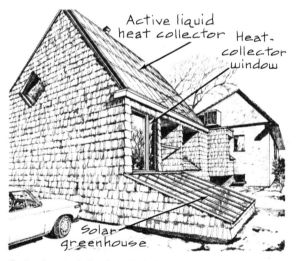

Solar-heated duplex: Unit in distance is an air-system retrofit; closer unit is a liquid system with three collector areas. Main collector-to-storage loop includes 560-square-foot double-glazed copper collector on house roof and 1000-gallon fiberglass tank in basement, with fan-coil system for heat distribution. Second collector is vertical window area in south wall, with bare, black-painted multi-finned heat exchanger mounted behind it as collector. For heating, water circulates up from 380-gallon black, uninsulated tank beneath window into heat exchanger. Tank itself is located in partially underground south greenhouse, which forms third passive heat collector. Architect: Richard Crowther.

A Variation on the Trickle-type Collector

The trickle-type collectors that form the main collector area of this house are unusual in their construction. Instead of a single corrugated absorber plate with water flowing openly over it, these collectors have two: one nested into the other, with a tiny space (1/32 inch) between them. The heat-transfer water flows through this space, in contact with both the upper and lower absorber plates throughout its passage from the pipes at the top to the outlet manifold which gathers it, heated, at the bottom of the collector array.

The system proves a neat solution to a problem facing most trickle-type collectors: it allows the water ample contact with the heat-collecting upper sheet (which is painted black and covered with double glazing) but prevents the condensation that robs open-face trickle-type collectors of heat through evaporation and contact of the condensed water with the glazing. This variation of the trickle-type functions much like a closed, tube-type collector, but it involves a less demanding technology and is considerably less expensive.

A small pump cycles the water from collectors to storage tank and back (the collectors drain completely at sundown to prevent freezing). The tank is of poured concrete, waterproofed, well insulated, and buried in earth beneath the house; its top supports the floor of one bedroom. Any heat lost from the tank serves to warm the bedroom above. The rest of the house is warmed by storage water thermosiphoned (with an occasional assist from a pump) up through the radiant floor slab (concrete with plastic pipes embedded in it).

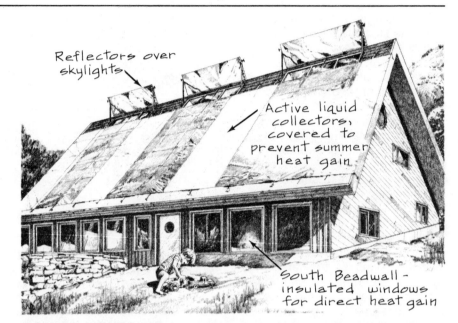

Solor heat collection occurs three ways in partially buried mountain home. Trickle-type collectors cover most of roof surface, facing south at 60° angle for winter sun. (In summer, canvas awning is tied tightly over collectors to mask them from sun.) Above collectors are three skylights, equipped inside with Skylid insulation and outside with reflectors, mounted on roof peak to bounce 25% more direct sun in skylights in winter, increase indirect light in summer. South windows allow passive direct heat gain in winter; overhang shades them in summer. Designer: Ron Shore.

Inside, house is open and airy; upstairs is a loft. Heat circulates freely from radiant floor heating system. Wood stove gives all necessary backup heat; its flue runs a long way up through room before venting outside, allowing heat to disperse to room air. South direct-gain windows are insulated with Beadwall at night.

Solar Assists These Heat Pumps

In most climates, neither a heat pump (see page 32) nor a solar heating system can, by itself, provide all the heat a house needs.

Put the two together, and the heat pump can plumb the solar storage tank for heat, even when the tank's liquid is much too cool for direct solar heating and the air outside is too cold for use as a heat source for the heat pump.

There are some tricks to making the combination economical, though, because the combined initial cost of the solar system and the heat pump can be considerable. The systems shown here have very dissimilar climates to contend with, but each solves the economy problem neatly, aided by high local fuel costs.

For the cold-climate house, the solar-assisted heat pump makes sense because for comfort the house needs a great deal of both heat in winter and cooling in hot, humid summers. The heat pump provides the cooling, and the solar system keeps the heat pump working efficiently during the heating season. The result: The house relies on its solar/heat pump team for almost 100 percent of its heating and a full 100 percent of its cooling needs, at a remarkable annual saving.

The mild-climate system, on the other hand, has neither great heating nor great cooling needs. So initial costs were kept down by using one of the most inexpensive solar systems: a low-temperature pool heater with plastic panels and the already-existing pool for heat storage. The system's efficiency was boosted by use of its heat-pump cooling cycle to add warmth to the pool during the swimming season by dumping unwanted house heat into it.

In freezing climate, an indoor solar storage tank acts as heat source for heat pump. 360 square feet of copper tube-type, double-glazed collectors surmount south-facing roof plane. Water heated in collectors keeps 5,000-gallon, heavily insulated poured-concrete storage tank in basement up to an average 70–100°F., supplying plenty of heat to heat pump. Backup is small electric water heater set to maintain storage water at 40°, plus a fireplace heat-recovery device through which storage water cycles. In summer, heat pump extracts heat from room air and uses it to heat domestic water, then dumps excess in storage tank. Designer: Spencer Dickinson/Solar Homes, Inc.

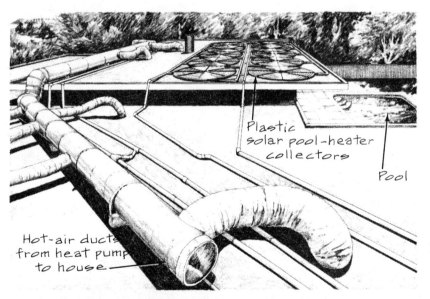

In mild climate, low-temperature solar swimming pool heater provides all assistance needed by heat pump. In winter, solar-warmed pool water is heat source for heat pump; heat is transferred to house via well-insulated ducting on roof. In summer, heat pump cools house by expelling house heat to pool. Designer: Paul Menyharth/The Solar Collector.

Placing Solar Panels When the Roof Aims Wrong

It's a common problem: You own a house and you're considering an active solar retrofit. But you check your house and discover that your roof isn't facing anywhere near south, or that it's so densely shaded as to be unusable as a collector area.

Don't give up hope! As you can see from the homes on these pages, there is a variety of options for collector location (also see page 40); one of them may suit your house and site's configuration.

Scout around and find out—as these homeowners did—where you have space on your lot for the collectors to face south without being shaded (the closer to the house the better, to minimize the complexity and heat loss of ducting or piping).

If you have a free, sunny south wall like that on the house at upper right, you might consider attaching a shedlike collector mounting to it. If your roof is flat, as is the one at lower right, your collectors can be placed on top of it, supported by a freestanding framework. If your house is aimed all wrong and the roof's tilt isn't correct, you might try angling the collectors off a deck area, sloping down below the house, as the owners of the home at lower far right have done. Or you can even mount them on a south-facing garage.

Of course, there are some homes whose lots are too small or shady, and whose orientations make solar retrofit impossible. But solar heating systems are much more adaptable than they seem at first glance, and adjustments in orientation, tilt, and size of collector area give surprisingly successful results with unpromising lots and homes.

Active retrofit liquid heat collectors

House looks north and west, so collectors were added to blank two-story south garage wall. Siding from wall was reused to cover framework that supports collectors and tilts them to greet winter sun. Collector framework also houses storage tank and backup heater; heat distribution is in garage. Designer: James Bukey/Solar Access, Inc.

Active retrofit liquid heat collectors

Flat roof and southeasterly orientation left homeowner/designer undaunted: he perched collectors atop roof, rotated them to face due south, and raised them on rack for correct solar exposure. Racks are set far enough apart that collectors don't shade one another. Shade trees drop leaves to expose collectors to winter sun. Designer: Alten Corp.

Active liquid heat collectors on garage

To let house face beautiful westward view, liquid-type collectors were mounted on south-facing garage roof. Underground distribution pipes carry heat to house. Designer: Alten Corp.

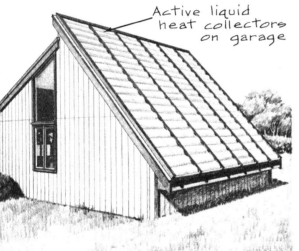

Active liquid heat collectors on garage

Bank of collectors below hillside home looks south while house looks southeast. Low-pitched roof and wrong orientation of ranch-style house made roof placement of collectors unwise, so foundation was extended to support collectors and hillside was excavated behind collectors to house a series of six small storage tanks plus all system controls. Broad-sweeping site insures no shading problems. Solar designer: Sunset Engineering.

Solar system comes with garage attached: prefabricated unit—garage plus liquid-type solar heating system—can be purchased for retrofitting, if your house is completely unable to adapt to a solar system. System is oriented on site as necessary, then hooked up to house. Design: Acorn Structures, Inc.

Active retrofit liquid heat collectors

If House Faces South, Adding Solar Is Easier

Active retrofit
air heat collectors

Graceful old mansion accommodates *its new solar heating easily. Air-system collectors were right at home on south-facing roof, tilted almost at optimum angle for solar collection. Inside, large closets were aligned conveniently to hold large air ducts, and rock storage bin was placed in capacious laundry room. Lack of unobstructed roof space kept collector area small: system heats only five of house's rooms, gives some spillover to adjoining rooms. Designer: Art Krill/Solaron.*

Active
retrofit liquid
heat collectors

The active solar retrofit problems facing a homeowner whose house already faces south and has plenty of unshaded roof and/or wall space are much less demanding than those for a homeowner whose house faces another direction.

If you've won the orientation battle, you must still confront three basic questions, however: Does the collector area tilt at the appropriate angle for solar collection at your latitude? (And if not, is there room for more collector area to make up the difference in quantity of heat collection?) Do you have room for the storage tank or rock bin and the distribution equipment? And is your house heat-conserving enough to make solar heating viable and economical? Three of the houses on these pages were retrofits; each had to tackle one of the above questions.

The house at lower right, for example, faced south and had plenty of room for the storage tank in the basement and excellent energy-tight construction. Its drawback was its roof angle: too shallow for good winter heat collection. The collectors had to be propped on racks to face the low sun.

Another house, at lower left, had a good roof for the collector

Close enough to south *exposure and desired tilt, this house's A-peak roof took collectors nicely. Liquid system was adapted to existing radiant heating; large storage tank was buried in yard, since space was lacking in house. Designer: Alten Corp.*

area (though the collectors had to be divided between two roof spaces), but no space adequate for the storage tank inside the house. The solution was to bury the tank in the yard.

Still another house, at upper left, had a roof with the right tilt angle and an orientation within 15° of south, as well as space enough inside for the air system's components. The catch was the house's ability to conserve heat. A big, rambling structure, it didn't have a great deal of insulation, and the old leaded windows were much too expensive to double glaze. So the system—a fairly small one—simply provides heat to a study, a workroom, and three bedrooms. The rest of the house is largely conventionally heated.

The last house shown here has none of these problems. This is because, even though its owners were not yet ready to invest in the solar system itself, they designed their house with solar in mind, so that when they can afford the system, the house will be very easy to retrofit.

Newly built home *is not yet retrofitted for solar heating, but foresighted owners had it designed as solar house, so future retrofit will be easy. South-facing roof angles toward winter sun; windows are concentrated on south wall; houses's interior is open, with lofts, to aid heat distribution; insulation and double glazing are excellent. Architect: Richard Ridley.*

One unshaded flank *of U-shaped house faces south, but roof pitch was too low to mount collectors flush with surface. To obtain optimal heat-gathering performance from collectors, each was lifted on a supporting rack to meet winter sun head-on, then braced firmly for protection against wind damage. Architect: David Waldron.*

Absorption Chiller Cools Hot-desert Home

Year-round hot weather combined with consistently bright sunshine prompted the architect of this house to make cooling the major objective of the home's design. He began by using some passive techniques for evading the sun to avoid heat gain (reversing the approach solar designers take in cold climates, where heat gain is the desired goal).

First, he oriented the house so its basic U-shape turned away from the sun to embrace (and shade) a northside courtyard and swimming pool. Even the north-facing windows and patio doors are deeply recessed for shade.

Then he planned a minimum of south-facing openings in the house—just a few small windows, protected from the sun by deep embrasures, slatted overhangs, and well-positioned trees.

Finally, he designed the house's low, sloping southeast and southwest profile to discourage the sun. Since the berms could not touch the walls (a restriction imposed by the mortgage company) for extra thermal protection, the architect shaped them to form a moat of shade around the perimeter of the house.

Having reduced the house's cooling load to a minimum, the architect added his most significant cooling effort: a full-size active solar-assisted absorption cooling system. Three tiers of flat-plate collectors on the house's south roof supply high-temperature water to the absorption chiller equipment inside the house, to activate the device's literal refrigeration of the house air.

Near-horizontal collectors, tilted back to face summer sun during height of cooling season, occupy most of south roof. They power absorption chiller in house. Below, on house's cool, shadowy north side, doors and windows open out to shaded courtyard, pool. Sun strikes courtyard only briefly on early summer mornings and late afternoons.

Low south profile minimizes sun's direct impact on this desert home. With year-round cooling as goal, south walls are virtually windowless, with rare openings protected like battlements by shade-casting buttresses and overhangs. High berms were planned as final sun-buffer, but mortgage company was too uneasy to support unfamiliar design concept, so earth was mounded away from walls for extra shading. Architect: Charles Martin.

Active Water Heaters ... described on page 29.

New Homes Take Well to Solar Water Heaters

As with any other kind of solar heating, solar water heaters are at an advantage when planned for a new home, because the design can incorporate them efficiently.

In the passive house at lower right, for example, the water heater's design benefited from the house's design flexibility, and vice versa, to their mutual advantage. The dormer on which the collectors are mounted lifts the roofline of the house, allowing extra headroom in the second-floor family room behind it. The heavily insulated water-heater storage tank was positioned against the far wall of the family room, close to the collectors to insure efficiency of heat transfer, but also near the upstairs bathroom and directly above the kitchen, again allowing short piping lengths and a brief tap warmup period. Pipes from collectors to storage travel up through the dormer roof and down the inner wall between the family room and the bathroom, so any heat loss goes to the house, not the outdoors.

The house with the trellis was designed to use its efficient copper collectors for both space and water heating: solar-warmed water goes first to the solar tanks, then to the gas-powered water heater, which boosts the heat, if necessary. The hot water may then be drawn off for showers, laundry, and dishwashing, or it may be sent through a fan-coil device or baseboard radiators to heat the upstairs rooms of the house as required. At this point, the architect/home-owner concedes that he cannot expect to shower while solar-heating the bedrooms; but eventually he hopes to supply all his heat with a larger solar system.

Unobtrusive solar collectors for active water heating are integrated with roof of home; trellis shades windows below. Four double-glazed copper panels, totalling 80 square feet, plus 30-gallon and 120-gallon solar storage tanks provide partial space heating as well as hot water. Misty, mild ocean climate obviates need for freeze protection, though system can drain down if necessary. Collectors heat efficiently even on foggy days. Architect: James Ellmore.

Active solar system heats household water for passive solar house. Roof was designed so dormer lifts three copper water-heater collectors to 55° angle for best winter sun exposure, while passive skylight heat-collector (in rear, with reflector/insulation panel raised) tilts back, admitting direct sun to heat rear-wall water-column storage mass in room below. Water heating system is thermostatically controlled; for protection against freezing, it pulses hot water back to collectors periodically on cold nights. House design: Jon Hammond and Jim Plumb/Living Systems; water system: David Springer/Natural Heating Systems.

Active Water Heaters: First-class for Retrofits

Older houses offer many obstacles to solar heating retrofits, but active water heating systems are just compact and adaptable enough to squeeze into some very uncompromising situations.

One of the two retrofits shown here (at lower right) includes an indoor hot-tub/whirlpool as well as the water heater. Water from the collectors on the roof above has arrived in the house as hot as 140°F. However, an installation error caused some early problems: due to a misplaced sensor, the differential thermostat gave no signals to turn the pump off. Consequently, water continued to circulate through the panels at night, giving off much of the heat accumulated during the day, and the backup electric heater was forced to take over. A simple repositioning of the sensor corrected the problem.

The second retrofit presented more design obstacles, but the installation went smoothly and the system performs well (see upper right). Since the slanting roof offered no collector area, the panels were mounted on freestanding brackets on the flat roof peak, tilted at 45° from the horizontal and rotated to face true south.

The house was being remodeled at the time of the solar installation, making it easy to run pipes through the yet-unfinished walls, down to the storage tank.

This particular water heater was sold as a kit, including panels and 82-gallon tank complete with prepiping and insulation, mounting brackets for the collectors, and an instruction manual. An experienced handyman could install the kit himself, but to avoid risks these homeowners hired the manufacturers to do the installation.

Active solar water-heater collectors

Perched on racks atop small flat rectangle at roof's peak, two collectors were positioned aloft because house roof had two drawbacks: it's oriented southwest, away from direct winter sun, and dormers crop out from every sloping roof face, disrupting potential collector area. Well-insulated solar storage tank was placed outside against house's west wall, because of space problems indoors; a redwood shed encloses it. Existing water heater for backup is inside house. Designer: Alten Corp.

Active solar water-heater collectors

Solar heating for a Victorian house: two collector panels, 4 by 10 feet each, found a suitable roost on sloping south roof of old wooden two-story home. Despite house's age, owner had no difficulty obtaining building permit for solar water-and-hot-tub heater, apart from a few questions on house's existing plumbing. Designer: The Solar Center.

For More Details— A Bibliography

Overviews

Daniels, Farrington. *Direct Use of the Sun's Energy.* New York: Ballantine Books, 1974.

Shurcliff, William A. *Solar Heated Buildings of North America: 120 Outstanding Examples.* Church Hill, N.H.: Brick House Publishing Co., 1978.

Williams, J. Richard. *Solar Energy Technology and Applications.* Ann Arbor: Ann Arbor Science Publishers, 1974.

Conservation and Climate Design

Aronin, J. *Climate and Architecture.* New York: Reinhold Publishing Co., 1953 (out of print).

Eccli, Eugene, ed. *Low-Cost, Energy-Efficient Shelter—for the Owner and Builder.* Emmaus, Pa.: Rodale Press, 1976.

"The Climate Controlled House." *House Beautiful,* series of monthly articles, October 1949 to January 1951.

In the Bank . . . Or Up the Chimney? A Dollars and Cents Guide to Energy-Saving Home Improvements. Washington, D.C.: U.S. Government Printing Office, Stock #023-000-00297-3.

Olgyay, Victor. *Design with Climate: Bioclimatic Approach to Architectural Regionalism.* Princeton, N.J.: Princeton University Press, 1963.

Shurcliff, William A. *Thermal Shutters and Shades.* Cambridge, Mass.: William A. Shurcliff, 1978.

Sunset Books. *Do-It-Yourself Insulation and Weatherstripping for Year-Round Energy Saving.* Menlo Park, Ca.: Sunset Books, 1978.

Villecco, Marguerite, ed. *Energy Conservation in Building Design.* Washington, D.C.: American Institute of Architects Research Corp., 1974.

Solar Design

Anderson, Bruce. *Solar Energy: Fundamentals in Building Design.* New York: McGraw-Hill Book Co., 1977.

Anderson, Bruce, and Riordan, Michael. *The Solar Home Book: Heating, Cooling, and Designing with the Sun.* Church Hill, N.H.: Cheshire Books, 1976.

Arizona State University, College of Architecture. *Solar Oriented Architecture.* Washington, D.C.: American Institute of Architects Research Corp., 1975.

Barber, E. M., Jr., and Watson, Donald. *Design Criteria for Solar-Heated Buildings.* Guilford, Conn.: Sunworks, Inc., 1975.

Crowther, Richard L., et al. *Sun/Earth: How to Apply Free Energy Sources to Our Homes and Buildings.* Denver: Crowther Solar Group, 1976.

Daniels, George. *Solar Homes and Sun Heating.* New York: Harper and Row, 1976.

Davis, A. J., and Schubert, R. P. *Alternative Natural Energy Sources in Building Design.* New York: Van Nostrand Reinhold Co., 1977.

Duffie, John A., and Beckman, William A. *Solar Energy Thermal Processes.* New York: John Wiley and Sons, Inc., 1974.

Jordan, A. C., and Liu, B. Y. *Applications of Solar Energy for Heating and Cooling of Buildings.* New York: American Society of Heating, Refrigerating, and Air-conditioning Engineers, 1977.

Kreider, Jan F., and Kreith, Frank. *Solar Heating and Cooling: Engineering, Practical Design, and Economics.* Washington, D.C.: Hemisphere Publishing Co., 1976.

Leckie, Jim, et al. *Other Homes and Garbage: Designs for Self-sufficient Living.* San Francisco: Sierra Club Books, 1975.

Lucas, Ted. *How to Build a Solar Water Heater.* Pasadena: Ward Ritchie, 1975.

Popular Science Solar Energy Handbook 1978. New York: Times Mirror Magazines, 1978.

Solar Dwelling Design Concepts. Washington, D.C.: American Institute of Architects Research Corp., 1976.

Total Environmental Action. *Solar Energy Home Design in Four Climates.* Church Hill, N.H.: Total Environmental Action, 1975.

Wade, Alex, and Ewenstein, Neal. *30 Energy-Efficient Houses . . . You Can Build.* Emmaus, Pa.: Rodale Press, 1977.

Watson, Donald. *Designing and Building a Solar House—Your Place in the Sun.* Charlotte, Vt.: Garden Way Publishing, 1977.

Wright, David. *Natural Solar Architecture: a Passive Primer.* New York: Van Nostrand Reinhold Co., 1978.

Yanda, Bill, and Fisher, Rick. *The Food and Heat Producing Solar Greenhouse: Design, Construction, and Operation.* Santa Fe: John Muir Publications, 1977.

Yellott, John I. *Solar Energy Utilization for Heating and Cooling.* Springfield, Va.: National Technical Information Service, U.S. Dept. of Commerce, 1974.

Solar Design Tools

ASHRAE. *Handbook of Fundamentals.* New York: American Society of Heating, Refrigerating, and Air-conditioning Engineers, 1977.

Climatic Atlas of the United States. Washington, D.C.: U.S. Government Printing Office.

Libbey-Owens-Ford. *Sun Angle Calculator.* Toledo, Ohio: Libbey-Owens-Ford Co., 1974.

National Bureau of Standards. *Intermediate Minimum Property Standards for Solar Heating and Domestic Hot Water Systems.* Washington, D.C.: Solar Energy Program of the National Bureau of Standards, 1976.

Uniform Solar Energy Code. Los Angeles: International Association of Plumbing and Mechanical Officials.

Solar Journals and Catalogs

Informal Directory of the Organizations and People Involved in the Solar Heating of Buildings. William A. Shurcliff, 19 Appleton St., Cambridge, MA 02138.

Solar Age. Solarvision, Inc., Church Hill, Harrisville, NH 03450.

Solar Age Catalog. Solarvision, Inc., Church Hill, Harrisville, NH 03450.

Solar Engineering. Solar Engineering Publishers, Inc., 8435 N. Stemmons Freeway, Suite 880, Dallas, TX 75247.

Solar Energy Associations and Information Centers

International Solar Energy Society. American Section: American Technological University, Box 1416, Killeen, TX 76541. (Write for a list of regional solar energy associations.)

National Solar Heating and Cooling Information Center. P.O. Box 1607, Rockville, MD 20850. Toll-free phone number: (800) 523-2929 (in Pennsylvania, call (800) 462-4983).

Solar Energy Industries Association, Inc. 1001 Connecticut Ave. NW, Suite 800, Washington, DC 20036.

Untangling Solar Terms—A Glossary

absorber plate: a black surface that absorbs solar radiation and converts it to heat; a component of a flat-plate collector

absorptance: ratio of solar radiation absorbed by a surface to the amount that strikes it (an important aspect of collector efficiency)

absorption chilling, solar-assisted: an air-conditioning method that uses solar-heated liquid to activate chilling process

active system: a solar heating and/or cooling system using mechanical methods of heat distribution

air-type collector: a solar heat collector designed to use air as heat-transfer fluid

berm: a mound of earth either abutting a house wall to help stabilize temperature inside house, or positioned to deflect wind from house

Btu or British thermal unit: basic heat measurement, equivalent to amount of heat needed to raise 1 pound of water 1° Fahrenheit

clerestory: vertical window placed high in wall near eaves, used for light, heat-gain, and ventilation

closed-loop: system in which heat-transfer liquid from collectors circulates through a heat exchanger immersed in heat-storage liquid, passing its heat to heat-storage liquid while remaining isolated from it

coefficient of performance: ratio of heat output to energy use of a heating or cooling device

collection: the act of trapping solar radiation and converting it to heat

collector: any of a variety of devices used to absorb solar radiation and convert it to heat

collector efficiency: ratio of collector heat output to amount of solar radiation that strikes collector aperture

concentrating collector: device that uses reflectors to concentrate direct solar rays onto a narrow absorber pipe to produce intense heat

convection, natural: heat transfer through a fluid (such as air or liquid) by currents resulting from the natural fall of heavier, cool fluid and rise of lighter, warm fluid

cooling season: portion of year (usually June to September) when outdoor heat makes indoor cooling desirable to maintain comfort

degree-day: unit representing 1° deviation of 1 day's mean outside temperature from a fixed standard (65°F.); used in estimating a house's heating or cooling requirements

design temperature: a designated temperature close to the most severe winter or summer temperature extremes of a climate, used in estimating a house's heating or cooling requirements

differential thermostat: automatic device that responds to temperature differences (between collectors and heat storage, for example) in regulating operation of an active solar system

diffuse radiation: solar radiation, scattered as it passes through atmospheric molecules, water vapor, dust, and other particles, so that it appears to come from entire sky, as on a hazy or overcast day

direct-gain system: passive solar heating system in which sun penetrates and warms house interior directly

direct radiation: radiation that comes directly from sun itself, capable of casting a shadow

drain-down: function of an open-loop solar water system in which all water drains out of the collectors when a freeze threatens

emittance: a measure of the ability of a material to give off thermal radiation (an important aspect of collector efficiency)

eutectic salts: substances that melt readily at low temperatures (as low as 80° to 90°F.) and, in so doing, store large quantities of latent heat, which they release when cooling and resolidifying

evaporative cooling: evaporating water cools and humidifies surrounding air; house air is circulated over water as a technique to cool indoor air in dry-climate areas

flat black paint: nonglossy paint with a relatively high absorptance

flat-plate collector: device that employs a planar absorber plate to collect solar radiation and convert it to heat, without assistance of devices to concentrate sun's rays

flow rate: the pounds of heat-transfer fluid which pass over or through an absorber plate per hour

Freon: a volatile chemical substance capable of boiling (becoming lighter) at low temperatures

"greenhouse effect": ability of glass or clear plastic to transmit short-wave solar radiation into a room or collector, but to trap long-wave heat emitted by room or collector interior

heat distribution: the act of conveying solar heat from collectors to storage and from storage (or collectors) to areas of the house where heat is needed

heat exchanger: device consisting of a long coil of metal pipe or a multifinned radiator, used to transfer heat from one fluid inside it to another outside it, without bringing the two fluids into direct contact

heating season: portion of year (usually October to May) when outdoor cold makes indoor heating necessary to maintain comfort

heat-recovery device: a device, designed for installation in a fireplace, through which house air or household water is cycled to reclaim fire's heat before it can escape up chimney

heat storage: medium that absorbs collected solar heat and holds it until it is needed to heat house interior

heat-transfer fluid: air or liquid used to carry solar heat from collectors to heat storage

hybrid system: solar heating system that combines active and passive techniques

indirect-gain system: passive solar heating system in which sun directly warms a heat storage element in one area of house, and heat is then distributed from that element to rest of house by natural convection, conduction, or radiation

insolation, or incident solar radiation: amount of direct, diffuse, and/or reflected solar radiation striking a given surface per hour

life-cycle costing: a method of cost analysis in which operating, maintenance, fuel, and other ownership costs are estimated for predicted lifetime of a device, and considered along with initial cost; often used to compare costs of solar heating or cooling systems and conventionally fueled systems

liquid-type collector: a solar heat collector designed to use a liquid as heat-transfer fluid

magnetic south: "south" as indicated by a compass; changes markedly from one location to another because of latitudinal relationship to Earth's magnetic fields

microclimate: climate of a very small area, such as a house site, formed by unique combination of topography, exposure, soil, and vegetation of site. Microclimate may contrast sharply with macroclimate (regional climate) in which it is situated

movable insulation: insulation placed over windows when needed to prevent heat loss or gain, and removed for light, view, venting, or heat

open-loop: system in which heat-transfer liquid from collectors feeds directly into heat-storage liquid

orientation: alignment of a building along a given axis to face a specific direction, such as along an east-west axis to face south

parabolic reflector: reflector designed in the shape of a parabola to focus extra sunlight onto absorber of a concentrating collector

passive system: a solar heating and/or cooling system using natural means of heat distribution—generally building's structure itself forms solar system

payback period: period of time a solar heating or cooling system takes to return its entire initial cost through fuel savings

percentage of possible sunshine: percentage of daylight hours during which direct sun is bright enough to cast a shadow

reflected radiation: solar radiation reflected off surrounding objects so it appears to come from them, as in reflection off a white wall or a car window

refrigerant: a volatile substance, such as ammonia, used for obtaining and maintaining low temperatures, as in a refrigerator

resistance, or R-value: capability of a substance to impede the flow of heat. Used to describe insulative properties of construction materials

resistance heating: a standard method of converting electricity into heat for the purpose of home heating

retrofit: to add a solar heating or cooling system to an existing home, previously conventionally heated and/or cooled

selective surface: specially adapted coating with high solar radiation absorptance and low thermal emittance, used on surface of an absorber plate to increase collector efficiency

sensor: device that detects changes in heat and relays information to differential thermostat

solar house: house that derives at least 40 to 50% of its annual heating (or cooling) from the sun

solar radiation: radiant energy emitted by the sun

space heating: heating of the air inside a building ("space cooling" is the converse)

stagnation temperature: high temperature range—300° to 400°F.—reached inside a collector on clear, sunny days when the heat-transfer fluid isn't circulating through the collector

storage mass: see "heat storage"

sun-tempering: designing a house to derive some of its heat directly from the sun (though not necessarily enough to qualify as a solar house)

temperature zones: areas controlled to maintain different temperatures within a house

thermistor: see "sensor"

thermosiphoning: see "convection, natural"

tilt angle: angle at which a flat-plate collector is tilted upward from horizontal for maximum solar exposure (maximum heat collection)

true south: south with reference to the stars, not to a compass; opposite to the Pole Star, which lies to the true north of Earth

Index

Flat-plate collector, 21–22, 24, 25, 30, 84. *See also* Glossary, 94–95
Freeze protection
  active liquid systems, 26–27
  roof ponds, 20, 62

Glossary, 94–95
Greenhouse, 18–19, 52–54, 56, 68
  retrofit, 39–40, 68
"Greenhouse effect," 9, 22. *See also* Glossary, 94–95

Heat distribution, 14–15, 22–24. *See also* Glossary, 94–95
  convection, natural. *See* Convection
  forced-air furnace, 27–28
  radiant baseboard, 27
  radiant floor, 27
Heat exchanger, 26–27, 28–29, 32. *See also* Glossary, 94–95
Heating demand, your house's, 35–36
Heating season, 35. *See also* Glossary, 94–95
Heat loss, 5–6
  analysis, 35–36
Heat pump, solar-assisted, 32–33, 85
Heat-recovery devices, 31. *See also* Glossary, 94–95
Heat storage. *See* Glossary, 94–95
  active, 23, 24–25, 27, 40
  passive, 14–21, 49–68
Heat-transfer fluid, 22, 26. *See also* Glossary, 94–95
Hot tubs, solar-heated, 30–31, 73
Household water heating. *See* Domestic water heating
Hybrid solar heating, 14, 28, 52, 53, 64–65, 77. *See also* Glossary, 94–95

Indirect-gain systems, 14, 17–21. *See also* Glossary, 94–95
Insolation, 36–37. *See also* Glossary, 94–95
Installing a solar system, 42–43
Insulating louvers, 12, 50
Insulating shutters, 11
Insulation, 6
  automatic movable, 12, 50, 55
  Beadwall, 12, 62, 84
  blown styrofoam, 12, 62, 84
  movable, 11–12. *See also* Glossary, 94–95
  reflective panels 11–12, 19–20, 38–39, 59, 60, 61, 69
  removable panel, 11
  Skylid, 12, 50
  sliding panel, 11

Landscaping for heat control, 13–14
Legal rights, 44–45
Life-cycle costing, 45–46. *See also* Glossary, 94–95
Liquid heat storage
  active, 27
  passive, 17, 19–20, 28, 58–62, 67
Liquid system, active, 25–28, 72, 80–85

Liquid-type collectors, active, 25–26, 83, 84. *See also* Glossary, 94–95
Loans, 47
Louvers, insulating, 12, 50

Mass, heat storage, 14–20, 48–68
Microclimate, 7. *See also* Glossary, 94–95
Mortgages, 46–47
Movable insulation, 11–12. *See also* Glossary, 94–95
Movable shading. *See* Shading
Multi-unit solar housing, 79, 83

National Solar Heating and Cooling Information Center, 46. *See also* Bibliography, 93
National Weather Records Center, 7, 37
Natural convection. *See* Convection, natural

Open-loop system, 26. *See also* Glossary, 94–95
Open-plan house, 9–10, 16
Orientation, 7–8, 38. *See also* Glossary, 94–95
Overhang
  fixed, 12–13
  hinged, 13

Parabolic reflector, 26. *See also* Glossary, 94–95
Passive solar heating systems, 14–21, 48–68. *See also* Glossary, 94–95
Percentage of possible sunshine, 36–37. *See also* Glossary, 94–95
Planning your solar system, guidelines for, 34–47
Pool
  covers, 30
  solar-heating, 29–30, 74–75
Preheating domestic water, 28–29, 67

Radiation, 5–6
Reflection problems, 45
Reflective insulation panel. *See* Insulation
Reflectors for vertical collectors, 38–39, 60
Retrofit. *See* Glossary, 94–95
  active, 23, 86–89
  home's potential for, 39–41
  ideas adaptable for, 48–92
  locating active collectors and storage, 40–41
  passive, 16, 68
Rock heat storage, active, 24
Roof angle for solar exposure, 38–39
Roof ponds, 19–20, 61–62
R-values, 15. *See also* Resistance *in* Glossary, 94–95

Selective surface, 21–22. *See also* Glossary, 94–95
Shading
  hazards, 38
  movable, 13, 69

Shutters, insulating, 11
Site for solar heating, 38
Skylid, 12, 50
Skylight insulation, 11–12
Sod roof, 10, 63–65
Solar heat collection, 9, 14, 16–26. *See also* Collection *in* Glossary, 94–95
Solar houses, 48–92. *See also* Glossary, 94–95
Solar radiation
  annual mean daily, 36
  available in your climate, 36–38
Solar system types, basic, 14–33
Spas, solar-heated, 30–31
Stagnation temperatures, 22, 42–43. *See also* Glossary, 94–95
Storage, heat. *See* Heat storage
Sun angles, 7–8
Sun rights, 44–45
Sunscreens, selective, 12–13, 52, 69

Tax incentives, 46
Temperature zones, 9, 16–23. *See also* Glossary, 94–95
Thermal mass. *See* Mass, heat storage
Thermosiphon. *See* Glossary, 94–95
  air collector, 20–21, 66
  water heating, 28–29
Tilt angles, 38–39. *See also* Glossary, 94–95
Trickle-type collector, active, 25, 84
Trombe walls, 17–18, 49, 55–57, 68
Tube-in-plate collector, active, 25
Tube-type collector, active, 25
Tube wall, 17, 58–59

Underground houses, 10–11, 63–65

Variances, 44

Water
  heaters, 28–29, 67, 91–92
  walls, 17, 58–60
Weatherstripping, 6
Windows, passive heat-gathering, 9, 11–12, 14–20, 48–51
Woodstove, backup heating, 31

Zones, temperature. *See* Temperature zones

Photographers

Holly Lyman Antolini: 72. **Edward Bigelow:** 49, 56, 57, 64, 74 top left, 80 top. **Steve W. Marley:** 50 top right and left, 55. **Norman A. Plate:** 50 bottom, 66, 79 top. **Joseph V. Saitta:** 80 bottom. **Solaron Corp.:** 79 bottom. **Werner Stoy, Camera Hawaii:** 71 bottom. **Darrow M. Watt:** 58, 63, 71 top. **Nikolay Zurek:** 65, 73, 74 right, bottom left.